Controlled Bombs and Guided Missiles of the World War II and Cold War Eras

An Inside Story of Research and Development Programs

Vernon R. Schmitt

Society of Automotive Engineers, Inc.
Warrendale, Pa.

For permission and licensing requests, contact:

SAE Permissions
400 Commonwealth Drive
Warrendale, PA 15096-0001 USA
E-mail: permissions@sae.org
Fax: 724-772-4028
Tel: 724-772-4891

Library of Congress Cataloging-in-Publication Data

Schmitt, Vernon R.
Controlled bombs and guided missiles of the World War II and Cold War eras : an inside story of research and development programs / Vernon R. Schmitt.
p. cm.
Includes bibliographical references and index.
ISBN 0-7680-0913-8
1. Guided bombs–History. 2. Azon bombs–History. 3. Guided missiles–United States–History–20th century. 4. Aeronautics, Military–Research–United States–History–20th century. I. Title.

UG1282.G8 S35 2002
623.4'519–dc21

2002066954

SAE International
400 Commonwealth Drive
Warrendale, PA 15096-0001 USA
E-mail: CustomerService@sae.org
Fax: 724-776-0790
Tel: 877-606-7323 (inside USA and Canada)
724-776-4970 (outside USA)

ISBN 0-7680-0913-8
SAE Order No. R-326
Printed in the United States of America.

Dedication

This book is dedicated to General Henry H. (Hap) Arnold for his vision and foresight during World War II in supporting Research and Development on guided bombs and missiles, which has not only given this country the military superiority over any other, but through the technology and technical advances these R&D programs created, made for a better life for everyone both at home and throughout the world.

Contents

Preface

The information set forth in this book is part historical and part designs that were included in our missile systems but were never completed. However, the basic technology for these systems was firmly established by R&D programs, the potential applications of which have never been fully explored. Advances early in the missile field came about mostly because of having a hands-on approach to the problems that were encountered. The advantage given to the ground forces by having ground control of bombs would have provided them with a devastating weapon for the invasion of Japan. The VB-3 would have been only a start; the Tarzon, a 12,000-lb bomb, was being developed and was ten times as powerful. A path could have been cut across Japan at least several miles wide with a minimum of ground troops. Such a tactic is not written as a competitor of the A-bomb but only to show we were making real progress in developing our offensive weapons.

The description of R&D efforts on the SLAM missile covers more detail because this weapon was undoubtedly one of the most destructive ever conceived, and in just a few years the basic technology to produce it had been developed. Fortunately for mankind, it was never built.

Acknowledgments

- The Vertical Bomb Group, Special Weapons Branch, deserve a very special thanks for the help they provided during the time we spent together on the development of guided bombs, 1944-45. They were as follows: Major J. Evans (Group Leader), Capt. William Hess, Lt. David Baltimore, and Cpl. Ralph Greenspan. Preparing bombs for test and on the test range, Cpl. Greenspan was a very capable and understanding worker, and was truly my right hand.

- A number of persons in the General Electric Co. contributed designs and tests of the Liquid Metal NaK-77 program; however, two names stand out: Dr. Hans Stern, who was responsible for the flueric servo valve design, and Robert Kumpitsch, Project Engineer and Manager of the program.

- Albert L. Pardini, author of *The Legendary Norden Bombsight*, furnished copies of Gulf Research and Development Company Progress Reports on their work on vertical bombs.

- Appendix A, "Pioneer to the Rescue," was taken from a story by Don Fortune in the *VFW Magazine*, October 2000 issue. It has been included here to demonstrate the devastation a guided bomb could cause. Don Fortune's book, *The Sinking of the Rohna, America's Worst Troopship Loss in World War II*, gives a very good description of this tragic event. Permission to use material from the magazine article and the book are greatly appreciated.

- Joe Baum, Battelle Memorial Institute, Columbus, Ohio, provided a number of recommendations on the General Electric NaK-77 program and was responsible for the metal bellows for the actuator that performed so successfully on the 1000°F test of the actuation subsystem.

- The United States Air Force Museum, Wright-Patterson Air Force Base, Ohio, provided a wealth of material for many of the photographs of airplanes and missiles.

- A very special thanks to Richard Kammerer, a co-worker in the Guided Missile Section of the Flight Control Laboratory, for his many suggestions during the preparation of this book and also his review of the data for validity and accuracy.
- It is impossible to heap too much praise on my wife, Betty, who typed the manuscript, made useful comments, and has put up with a cranky old "do-it-right" person for the past 55 years.

Chapter 1

Introduction

When David slew Goliath he used a missile. In his case the missile was one of five smooth stones he had gathered from a nearby brook. This story demonstrates how brainpower can overcome bulk and brawn by controlling a missile to a most vulnerable part of another human—his head.

Missiles may take a variety of shapes and forms; however, missiles and guided missiles as we apply the terms need to be defined in order to fully understand their design and use. As might be expected a number of definitions exist; hence, a listing of these plus comments will prove quite helpful.

Webster defines a missile in two parts:

Missile 1. A weapon or object thrown or projected
2. A self-propelled unmanned weapon (as a rocket)

Guided Missile A missile whose course may be altered during flight (as by a target-seeking device)

Obviously Webster's definition of missile is acceptable but the definition of a guided missile leaves much to be desired.

Two definitions used by the SAE (Society of Automotive Engineers) are more descriptive and complete in terms of today's designs and operational usage:

Guided Missile An unmanned vehicle moving above the surface of the Earth, whose trajectory or flight path is capable of being altered by an external or internal mechanism. (SAE Aerospace Recommended Practice)

Guided Missile An unmanned airborne vehicle whose flight path or trajectory can be altered by some mechanism within

> or attached to the vehicle in response to either a preprogrammed control sequence or a control sequence transmitted to the vehicle while in flight. (SAE Aerospace Dictionary)

From the above definitions one can readily understand how easy it is to conceive of the guided bomb becoming a guided missile.

The design and development of a guided missile depends on the intended use and purpose as a weapon. The term "weapon system" in most cases would be considered more appropriate by today's standards. However, if we just consider the missile, it is composed of a number of components that are common to all types, namely: (A) a warhead, (B) launching method, (C) guidance system, (D) propulsion system, (E) control system, and (F) target.

The warhead can be considered as part of the missile; in fact, in the early programs the controlling members were attached to the bomb (warhead). As the weapons system became more sophisticated the warhead was placed in the most convenient location, usually in the nose or forward section of the missile. The size and type of warhead was determined by the type of target and mission purpose.

Except for underwater launching there were for all practical purposes only two methods to choose from, air or ground. Air-launched missiles, like delivery of bombs, would require a stable platform for the launch. Ground-launched missiles could be launched from a trailer with the aid of a rocket, referred to as zero launch, or from a track. Other force-producing designs were either considered or used on several designs, all of which aided in bringing the missile up to flight speed where the main propulsion system could take over.

Guidance systems varied quite widely from inertial, radar map-matching, heat-seekers, and combinations of these and other methods, all of which were designed to steer the missile to the target.

There were a variety of propulsion systems employed. A number of the earlier missiles used aircraft engines and several used rocket types.

The control system provided the components and subsystem parts to keep the missile properly oriented and stable aerodynamically in order to stay on the trajectory defined by the guidance system. The control system basic components consisted of sensors, usually gyros for orientation, electronic computational networks, and a servo actuation system for moving aerodynamic control surfaces which, in turn, provided the forces required for stability and control.

In most cases the target was a stationary object such as a bridge or building, or an area of concentration such as a rail or troop center. However, on some systems the missile chased another missile or aircraft. In this case the "Curve of Pursuit" is applied, which means the chasing missile must have a higher velocity than the object it is chasing. Moreover, to complete the interception one must assume the chasing missile velocity always remains higher than the missile it is chasing, that the first remains on a relatively straight course, and that the chasing missile has the maneuverability to stay on an interception curve or path.

The use of aerial bombing in wartime had been considered for many years and, although having seen limited use in WWI, this had never been a real threat until WWII. The possibility of hitting a target by eyeball sighting from a great height is quite remote. Hence, bombsights and techniques for bombing, such as dive-bombing, were developed, providing the means to destroy a certain target or installation. However, there would be an added advantage and improved bombing accuracy if a system could be installed in the bomb that would permit guiding it after it was released.

Although work had been undertaken in Germany for the guiding of bombs in the late 1930s and continued during WWII, which included the V-1s and V-2s, in the United States such technology was not available at the start of WWII and had to be developed after 1940. Along with development came the necessary experimental work and tests. This took time and the number of individuals with the technical background required for this work was very limited. In fact, the evolution from design concept to equipment that could be used in actual warfare was, in most cases, still in the experimental phase when the war ended.

The bombs described in the following pages are intended to provide an understanding of the magnitude of work and the problems involved with the development of new weapons. From this early work came some of our more sophisticated missile systems that are currently a part of our military's operational equipment.

All vertical bombs, including missiles or guided missiles, have two common characteristics: they are released by airborne vehicles and powered to the target by gravity. As we examine the progress in the field of guided missiles in the U.S. from the early beginnings in WWII into the 1970s, one recognizes how many disciplines and integrated technologies were involved. The fields of aerodynamics, propulsion, and flight control were the ones that perhaps advanced the most rapidly. However, we must remember to include the use of nuclear power (for example, submarines that far outperformed the WWII models), and finally the development of a ram-jet nuclear power plant.

It would be difficult to measure, or even to estimate, the technological advances made in the United States without the development programs on guided missiles for the Cold War after the end of WWII. The severe environmental conditions that missiles were subjected to forced the development of better materials and designs to withstand these conditions and perform as required for their intended use or mission.

The list of improved designs would be quite extensive, but, just as examples, in electronics we moved from the vacuum tube to mag-amps and printed circuitry, then to solid-state electronics which not only could survive in the shock and vibration environment but also had a greater output and in turn led to direct-drive servo valves. In the flight control area came adaptive systems, then self-adaptive, that is, the system would adjust to changing flight and environmental conditions and in turn improve the reliability. Also with the adoption of electronics in the signal path was the use of redundancy and the fly-by-wire system that provided greater safety and reliability for aircraft.

When one adds all the advances that were used by the aerospace industry into making better and safer commercial flights it is no small wonder that the United States leads all other countries in air travel. Furthermore, one must keep in mind it was this same technology that was used to launch communication satellites, put a man on the moon, and lay the groundwork for other space travel now being contemplated.

The research conducted in the various disciplines to solve design problems on guided missiles and other weapons during the Cold War did more than just provide data. It demonstrated in the minds of most people that without research, or R&D efforts, little will ever happen to advance the knowledge and understanding of that discipline. For example, the various space programs eventually put a telescope in space, the Hubble Telescope, and the payoff has been next to phenomenal, removing some of the myth and superstition about our solar system and other heavenly bodies. It gave astronomy perhaps the greatest boost it has ever enjoyed. Time-wise, this research was conducted over a period of approximately 50 years. Without it we would have progressed, from a technical viewpoint, only two or three years in that 50-year span.

Without ever consciously recognizing it, the technology provided by research to win the Cold War was the greatest factor in this country gaining and maintaining the world leadership in scientific knowledge, including advances in the field of medicine.

Chapter 2

Vertical Bombs of WWII

U.S. Types

The Azon

The Azon bomb, or VB-1, was a radio-controlled bomb that was controlled in the azimuth direction, from whence it took its name. The tail assembly that housed the radio receiver, gyros, and electric motor to actuate the aerodynamic controls replaced the regular tail and was attached to the rear of a 1000-lb standard high-explosive bomb. The gyro assembly was comprised of two gyros: The directional gyro provided a reference for directional control and a rate gyro provided a rate control about this reference using ailerons that were solenoid-operated in bang-bang manner, giving an oscillation of about ±5 deg. Total amount of control from 20,000 ft. was estimated at 600-700 ft. assuming control was applied soon after release. However, it is understandable that control would be greater during the latter part of the trajectory as the bomb picked up speed and moved into atmosphere of a greater density. A pyrotechnic flare at the rear of the bomb, ignited approximately 6 seconds after release, gave a high-intensity light. This provided the bombardier or controller the position of the bomb relative to the target and its flight path. Since the bomb was controllable in azimuth, the usual procedure was to guide the bomb along a road, or imaginary line, that would intersect the target. Thus, some experience or practice on the part of the controller would prove most valuable.

The Azon bomb did see some limited use in the European, Mediterranean, and China, India-Burma Theaters during WWII. However, one of the useful attributes of having the Azon was that it provided a real springboard of data for the development of the Razon. There was some discussion in the early stages of having a VB-2 which would have been a souped-up version of the VB-1 on a 2000-lb bomb, but having the range control capability on a controlled bomb far exceeded the need for a higher explosive one. A number of Azon bombs were used for experimental purposes on other projects such as Spazon and ground control.

Instrumentation and recording of data were always a problem. We didn't have any telemetering equipment and hence obtaining in-flight data was a chore. Most of the Azon flight data was recorded on paper via a small six-channel brush recorder. Then in order to retrieve the paper recordings, the bombs had to be dug up from the bomb floor, usually under 12-16 ft. of sand. In short it took time to prepare what we were going to record and then added time to recover the data. It was in good condition when recovered since it had been packed in a sand-loaded bomb and could take the shock. When the bomb was dragged out of the ground, the back plate had to be removed in order to get to the recordings which had been put in a small paper box and placed near the rear of the bomb.

Figure 1. Photo of the Azon.

Figures 2a and 2b. Photos of tail assembly and Azon ready for installation in aircraft.

Demonstration of Azon

Our branch, The Special Weapons Branch, had a display room where a number of our special weapons were on display. One that was used to show the infrared sensor and control was called the VB-6. On the nose was mounted a heat-seeking sensor and when the demonstrator lit a cigarette lighter and moved it from side to side or up and down, the nose assembly would move in that direction. We had several Azon bomb tail assemblies on display but they didn't do anything. Thus a thought—"Why not an in-motion display of the Azon?"

Sketches were made for a display stand. The Azon bomb with tail assembly attached was installed on the modified bomb stand with the bomb rotational motion around the longitudinal axis, simulating roll. Rollers were added to the bottom of the stand and a motor mounted so that when signal for azimuth control was applied, the entire unit would rotate about the vertical axis that ran through the center of the bomb when directional control of the bomb was applied. An electric light was installed to simulate the flare, and a servo motor connected to a lower bomb bracket caused the bomb to oscillate approximately ±10 deg, and, thus, through the gyro activate the spoilers. A test stand with controllers was installed nearby and a simulated flight with controls could be demonstrated.

Colonel Holloman (Holloman Air Force Base), our laboratory chief, thought this to be quite a unique display and one day in early 1945 brought a visitor in to see it operate. His visitor was Orville Wright. Orville was quite interested in radio-controlled vehicles and made one or two visits a year to the laboratory. On this particular day there were only three persons present: Colonel Holloman did practically all the talking, Orville listened intently and smiled a number of times, and as instructed by Colonel Holloman, I operated the controls and demonstrated an Azon bomb mission. Everything worked as it should and I was thankful that my design and display was to obtain some notice after all. Later there were discussions about modifying this unit so it could be used as a test stand for Azon and Razon bombs but, as usual, projects with higher priority took over and it remained another display unit.

Ground Control of Azon

Knowing the background of how and why ground control of the Azon bomb came about is absolutely necessary if one is to appreciate the problems and solutions that were involved. Understanding the advantages of ground control is easy but providing and carrying out the implementation creates problems of the first magnitude. It was like moving the bombsight from the aircraft to the ground. Now the ground operator must guide the aircraft to the desired release point, release the bomb at the proper instant, then guide the bomb to the target.

In many cases an observer on the ground can see a target without difficulty at 2-3 miles, whereas a bombardier has to spot the target miles away sometimes through cloud cover and do all this in a short time.

In early 1945, as a member of the vertical bomb group at Special Weapons Branch, Wright Field, I wrote up a proposed bombing technique on control of vertical bombs. The proposed version used the VB-3 Razon bomb and the B-17 aircraft with ground station for azimuth control and another ground station approximately one mile from the first as a range station.

A team of weaponry experts who were stationed in Washington, D.C., and examined all new and proposed weapons reviewed this proposed technique of ground control on the VB-3 bomb and recommended and directed that a project be set up to determine the feasibility of employing it. A project was set up in our Group of the Special Weapons Branch and given a 1A priority. At this point one must recognize the need and potential use of this method. We had been losing marines by the hundreds, if not by the thousands, in the island-hopping in the Pacific, and this looked like a viable solution. The bombs would come in at a high angle and with plenty of time to hit targets 1-2 miles away. From the implementation point there was little data and much less in the way of equipment identification. Nobody had ever guided a bomb from the ground before and how did one bring the aircraft to the proper release point? What we didn't have was easier to set down than what we did have. We were short on time and personnel; i.e., we didn't have an engineering staff to draw on for the development of the sophisticated equipment similar to that currently being used for bombing.

Personnel-wise we had: a staff sergeant for project engineer who also had made the original proposal and his assistant, a corporal; a captain who took care of the administrative work securing and scheduling the program; and a lieutenant who handled the radios and electrical equipment. We did not have any sighting equipment or experience in handling the necessary ground equipment that would be required. In a sense we were starting from scratch.

The first item to be undertaken was the sighting device. A pictorial diagram of this is shown in Figure 3. It was a crude design by any standard; however, it did the hurry-up job. This was constructed at Wright Field and taken to Wendover where we did all the test work. In order to save time and determine just how feasible this would be, it was decided to use just the azimuth control or one ground station and the Azon bomb.

One must keep in mind that all the functions performed by the bombsight and electromechanical calculating system which accomplished this automatically must now be done manually by the ground station and whatever means they could devise. How many unknowns that occurred that were never solved remains unknown. However, all the basics were solved and the bombs dropped using what was available to the best of our ability. Our group of four worked as a team and, however things turned out, we shared failures and successes together.

Plans at the start of this project included use of the VB-3 Razon bomb and also the design and use of a range-sighting device which would be at another ground control station located a mile or two from the azimuth control station at an angle of approximately 90 deg. It was thought that at least four or five Azon bombs would be dropped in order to establish repeatability and reliability of equipment and experience for personnel. The range-sighting device would be designed and constructed during the Azon drop tests. This was to be accomplished in a time frame of two to three months, which appeared to some that we were moving too slowly. However, the real problem was that we just didn't have enough qualified personnel.

Our team did get together to continue work on the project in early August 1945. However, the A-bomb had been dropped and we were put on hold. Finally the war ended and no more research was performed on ground control of vertical bombs. Unfortunately, to my knowledge there has never been any follow-up work, and like so many projects, research died on this effort at the end of World War II.

Sighting Device and Bomb Steering

In a sense the sighting device was a miniature of the conditions that actually existed for dropping and guiding the bomb. Figure 3 is a pictorial view of the main section of the device. It had a piece of 3/16-in. plexiglass mounted on 3/4-in. plywood. The sides were about 24 in. apart and approximately 30 in. in height, leaving room for the controller with control box to stand inside as the upper section was mounted on a tripod stand. This structure could rotate about a pin in the front portion and thus align the sight line to the target. A bubble level was also installed to permit leveling of the sighting device. The plexiglass extended rearward about 2 ft. and had pencilled lines marked in the diagram as E, E′, etc., to aid the controller in communicating to the aircraft pilot instructions to null out the drift and guide the aircraft to the desired point of release. This took a minute or two and with this was instruction to check the bombs for warm-up and open the bomb-bay doors.

Information we received on this test run was that no one in the aircraft could see the target clearly because of a slight cloud cover. When the aircraft passed over the line shown in Figure 3 as AR, the point of release, the controller pushed the bomb release switch, which by radio control activated the bomb shackles and released the bombs—the controlled bomb and a dummy. Six seconds after release the flare came on. The bomb at this time was to the right of the thread and, hence, control was applied to move it to the planned trajectory. Drift angle at time of release was estimated at 5 deg. This was based on the amount of control applied soon after release to get the bomb to follow the thread that was the predicted trajectory. The bomb responded to control quite well but tended to drift to the left; however, as control was applied it responded in kind. Control was applied as necessary to keep the bomb moving along the thread which it did until impact.[1]

[1] A suggestion was made after the test that a high-intensity intermittent electric light be beamed in the direction of the aircraft to aid the pilot in locating the ground station prior to and during the bomb run.

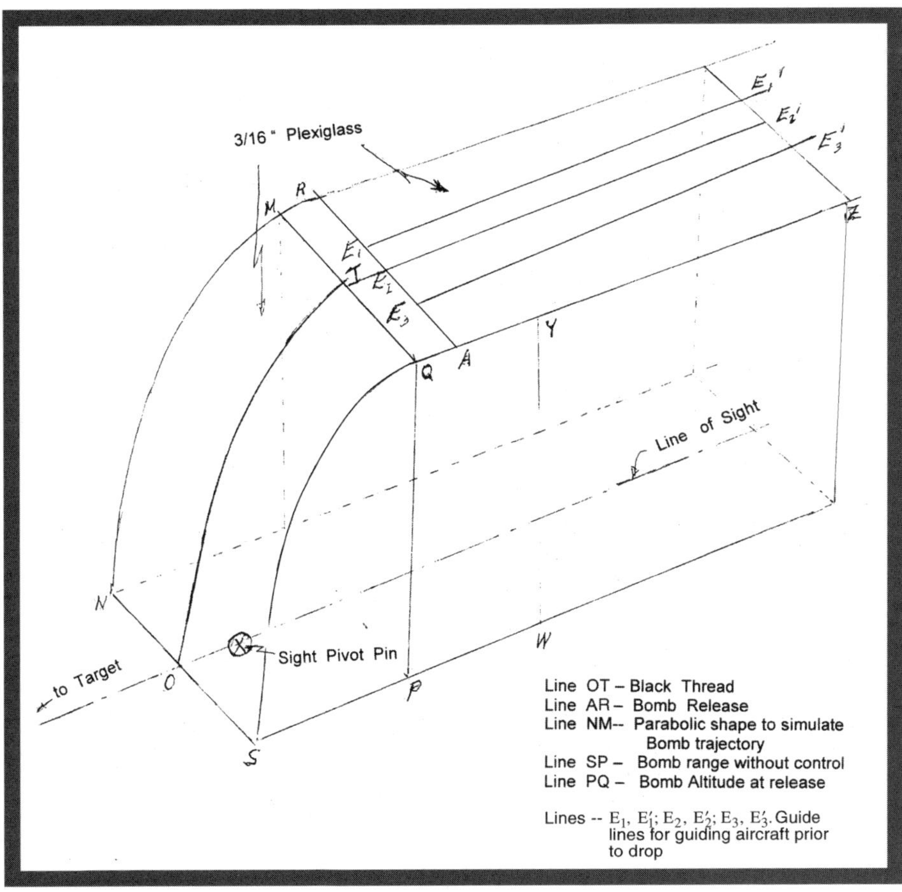

Figure 3. Ground control of Azon bomb sighting device.

Right after impact the controller took off to the target area. The range crew were quite elated; they said it missed by only 10 ft. in azimuth and landed about 150-200 ft. beyond the target. This demonstrated that the release point was fairly close to what it should have been. Impact point of the controlled bomb was easy to locate as it always left a smoldering area in the sand.

Immediately on returning to base, our group leader, Major Evans at Wright Field, was informed of the results. During the previous two weeks, a Marine Lt./Col. had been assigned to monitor our project, but after this successful drop he disappeared and we were informed that for the present there would be no further testing. Apparently the Marines wanted to employ this bombing technique right now and were ready to use Azon bombs. I was instructed to return to Wright Field and was given a two-week furlough. It looked like an overseas assignment was in the near future.

Ground Station

A plot for ground control of vertical bombs is shown in Figure 4. It is not to scale and intended primarily to give a sense of location, e.g., the outer range marker was a beacon located about 25 miles from target.

In setting up the ground station a number of assumptions were made as follows:

(a) We are in enemy territory and must stay at least 1-1/2 miles from the target.

(b) Our aircraft will be flying at a true air speed of 200 mph and bomb will be released from an altitude of 20,000 ft.

(c) We have complete and uninterrupted communications with the aircraft at all times during the bombing.

(d) Time of fall for controlled bomb was approximately 37 seconds. Once the ground station site was selected certain prerequisites had to be established:

   (1) Magnetic heading of sight line; this turned out to be 260 deg which in turn was sent to the pilot.

   (2) To determine our distance from the target we chose a point perpendicular to the sight line and a distance of 200 ft. from the ground station measuring the angle between this line and the target. This was 88.75 deg or distance to target = 200/tan1.25 = 9174 ft.

   (3) Release point range is 200 x 37 x 1.47 – 1500 = 9378, or release point is approximately 200 ft. away from target from a point directly overhead of ground station; thus the line shown as AR in Figure 3.

   (4) To determine that ground station is at least 1-1/2 miles from target, we have 9378 – 5280 x 1.5 or 9370 – 7920 = 1450 ft. on

the positive side. Assigned to the ground station was a truck that had power supplies, communications equipment, and radio transmitting equipment for sending signal to the bomb and release of bombs in the aircraft. In other words, the truck with equipment was part of our ground support equipment.

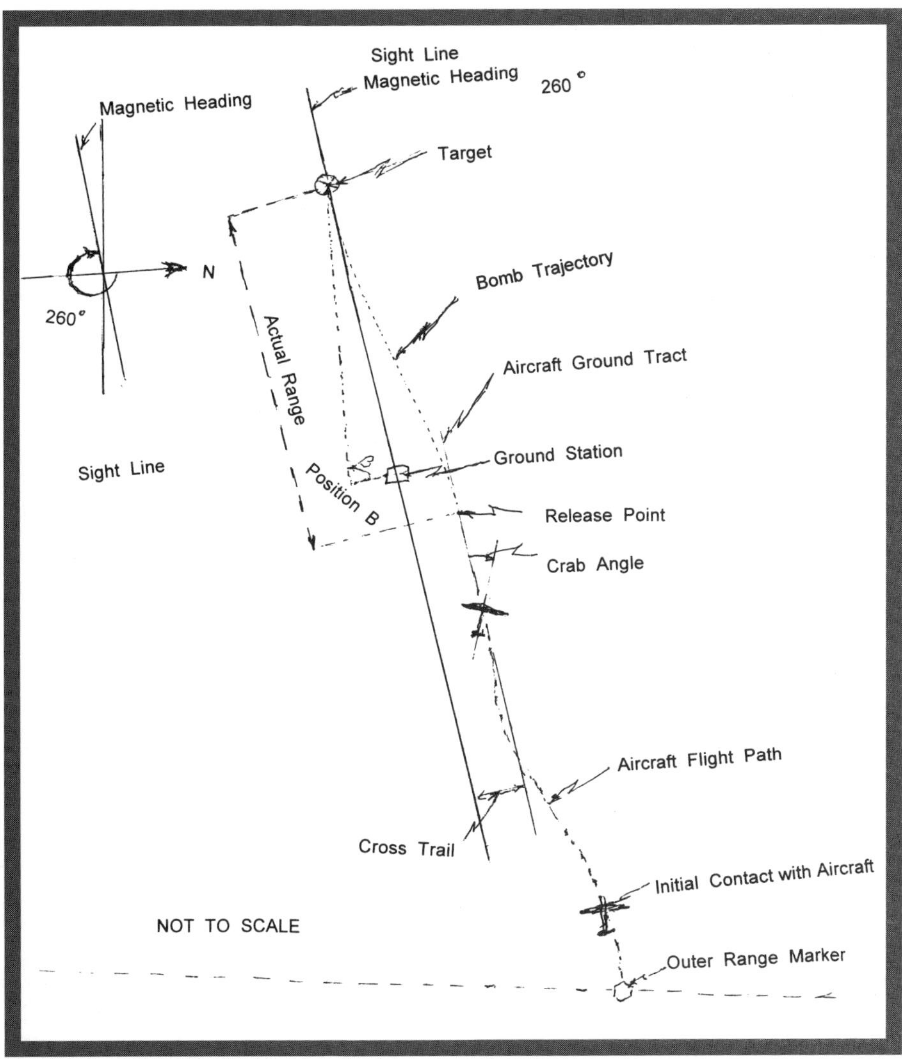

Figure 4. Plot for ground control of Azon.

Figure 5. Project Engineer Sgt. V. Schmitt and driver arrive at ground control station on bombing range, Wendover, Nevada. (June 1945)

Figure 6. Sgt. V. Schmitt and Cpl. R. Greenspan preparing bombs for VB-3 tail assembly. (Photo shows Cpl. Greenspan and Sgt. Schmitt preparing bombs prior to installing the tail assemblies that controlled the bombs in flight. In background is B-17 aircraft used in drop tests of the bombs.) (June 1945)

The Razon Bomb

In order to appreciate and understand the development and use of the VB-3 Razon bomb more completely it is desirable to divide its description into the developmental and operational phases.

Developmental Phase

More resources were allocated to the development and testing of the Razon than any other vertical bomb. Undoubtedly this was because of the data and experience provided from work on the Azon bomb and also because there was a ready operational need once the bomb had been test-proven. Photos of the Razon side and rear views are shown in Figures 7 and 8.

Figure 7. Side view of VB-3.

Figure 8. Rear view of VB-3.

The Gulf Research and Development Company under Dr. Wycoff's direction and under an NDRC (National Defense Research Council) contract was responsible for the development and test of the VB-3. The Vertical Bomb Group, Special Weapons Branch at Wright Field provided military personnel who assisted and supported Gulf in their work. This covered the period from November 1944 to September 1945.

During the developmental phase, sand-filled 1000-lb bomb casings were used with the VB-3 tail assembly attached to the bomb. The rear shroud called the control shroud had moveable surfaces which responded to control inputs for pitch and yaw or up-down and right-left. The displacement of these surfaces in flight produces aerodynamic forces on the bomb causing it to rotate from the reference provided by the directional gyro. In front of the control shroud was the "lift" shroud that had been added to furnish a forward force when increased range was needed. The lift shroud would have been more effective if located nearer to the c.g. (center of gravity) but other factors made it a part of the tail assembly, which in terms of weight was approximately 185 lb. Drop tests of the VB-3 from November 1944 to September 1945 were made from a B-17 at Wendover Air Force Base, Utah.

As a radio control bomb the VB-3 was steered by the bombardier after release and his inputs were in response to how well the bomb followed the desired trajectory. A pyrotechnic flare attached to the rear of the bomb came on shortly after release and provided the bombardier the relative position of the bomb to the target.

Ninety-three VB-3 bombs were used in the Gulf test phase, the purpose of which was to determine operating data on the Crab and Jag bombsight attachments. Twenty-five tail assemblies were built by Gulf and had aluminum shrouds giving a total weight of the tail assembly of 166 lb. The remaining 68 tail assemblies were built as production models by Union Switch and Signal Company, Swissvale, Pennsylvania. These assemblies had steel shrouds which gave a tail weight of 184 lb. Figure 9 gives the dimensional locations and aspects of the bomb components. The anemometer fuse and flare arming device were also tested during these drop tests. Operational data on the JAG attachment obtained on these tests proved that it would increase the overall accuracy of the VB-3 immeasurably. The JAG permitted the added time of fall of the bomb due to its maneuvering to alter bombsight settings and thus synchronized the computed time with the real time. This was not a critical factor for azimuth control but for range it had to be taken into account.

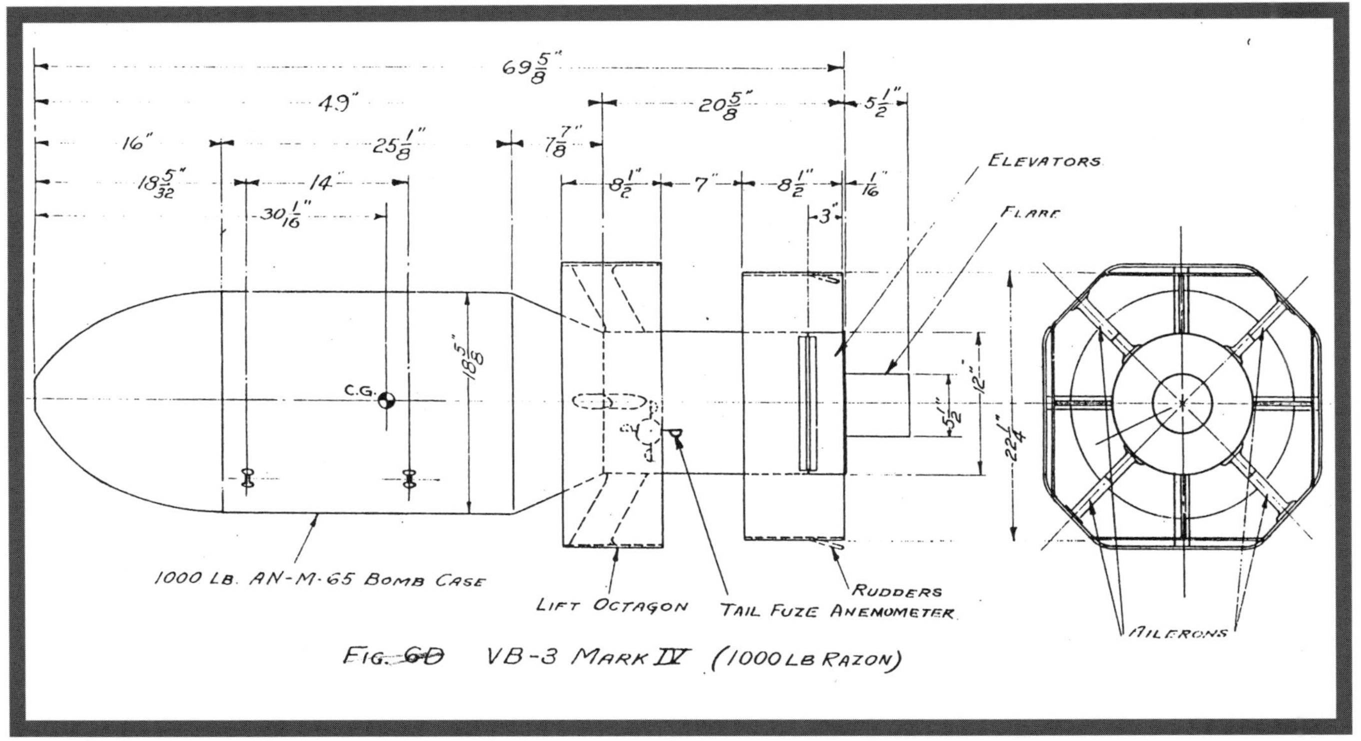

Figure 9. Dimensions of the VB-3.

In their report, Gulf provided the following failure rate data on the 93 VB-3 bombs dropped during this test phase:

(1)	Failure of the JAG to operate or to be clutched in	5 bombs
(2)	Inadequacy of control for correcting a very large aiming error	3 bombs
*(3)	Disappearance of flare image in the field of view of bombsight	5 bombs
(4)	Bombardier error in use of incorrect value of disc speed	1 bomb
(5)	Rack failure causing late or early release of bomb	2 bombs
(6)	Yaw angle test early in flight upset the steering accuracy which immediately follows	2 bombs
	Total	18 bombs

* The loss of flare image may have been due to an unstable bomb, which was observed by a member of the Bombing Range Crew. An account of what was observed is included below, titled "A Razon bomb goes unstable."

This failure rate is about 1 in 5, which was about the same as occurred during operational use.

With regard to the change in time of fall, the Gulf report gives this statement: "the probable variation in time of fall about the average was ±0.53 second at 10,000 ft. altitude, ±0.49 second at 15,000 ft., and ±0.85 second at 20,000 ft."

Further development work on the Razon bomb was discontinued after WWII.

A Razon Bomb Goes Unstable, June 1945

On this particular day I decided I would observe the bombs from the target area. The day was clear and with very little wind—it was a day for test work. Two sets of Razon bombs were to be dropped from an altitude of 20,000 ft.; i.e., altitude above target. The aircraft was a B-17 and all flight and test personnel were experienced individuals. I had helped load these bombs the night before and felt equipment-wise everything was in order. It was mid-morning when the test aircraft appeared over the range. After another 10 minutes I heard over our system "bombs away." I could see the aircraft but not the bombs. Then approximately 6-7 seconds later on came the flare. I did the usual, i.e., finger track the bomb to see if it was coming in my direction. I noted it was moving off to the left so I was OK. I took my binoculars and trained my sighting to the flare. The bomb was in view and I could even make out the details. I don't know if the controller had made any corrections previously but I did detect a slight oscillation in yaw which I guessed to be a cycle per 2 to 2-1/2 seconds with an amplitude of 5-10 deg. There was also a slight motion in pitch of perhaps ±5 deg. After another 5-10 seconds, now 15 seconds in its flight, I noticed the yawing increasing probably to ±10-15 deg and there was a tendency to slow the motion near the end of its travel. I was fascinated as we had never had any recorded flight data of this type. Finally after several more increased yaw motions, it failed to return to oscillatory motion and went into a flat spin of approximately one cycle every 2-3 seconds. The spin rate seemed to increase as the bomb fell to about one cycle a second. We were now into a flight time of some 20-25 seconds. I just could not take my binoculars off this bomb. I also noticed a slight tilting of the flat spin; i.e., it now took on another oscillatory motion in pitch of say 5-10 deg with each revolution. Then it suddenly went into a tumbling motion, the flare still burning and the bomb turning end over end. This continued until impact which was about 200-300 yds from my observation position.

In the past we had a number of bombardiers and bomb controllers report that occasionally a bomb would turn or tumble but had no recorded data. This happened in about 5-10% of the flight tests. I didn't write up my observations in our report but I did pass the word along to others just as I have described. My evaluation of the problem was that the instability was caused by aerodynamic conditions and that we needed to add some aerodynamic damping in yaw. I also suggested we get some wind tunnel tests on both the cause and the design used to solve this problem. I was convinced that we had an aerodynamic problem rather than a control problem with our current design. The war ended before we solved this problem or conditions that caused the problem.

Operational Use of the VB-3 Razon Bombs in Korea

Early in the Korean War a decision was made to use the VB-3 bomb in Korea.

Near the end of WWII in 1945 a shipment of VB-3s was made to the China-India-Burma theater. They were stored in a warehouse and apparently stayed in the "as received" condition for the next five years, never unpacked or inspected during this storage time. A squadron of B-29s stationed at Okinawa needed to fly these bombs to targets in Korea. This shipment of bombs that had been stored for over five years was sent to Okinawa and had to be put in operational condition.

If it can be assumed that the development work on the VB-3 was completed by the Gulf Research and Development Company on their contract in 1944, then the tail assemblies that were built after should have been production models and ready for operational use. Of course this would have been right after completing development and not after a long storage time. These tail assemblies were in very poor shape when received at Okinawa. The bomb tails came boxed with the components packed separately but all wrapped in supposed-to-be barrier material. However, due to poor storage conditions, or shipping and handling, about one in five had been broken and exposed to the elements. Many components and some parts of tail assemblies were rusty or covered with mold, and several even had ant nests with ants still remaining when unpacked. As tail assemblies were unpacked and inspected it seemed like we were on a salvage operation. We never kept an official count of the bombs we uncrated as opposed to the number we actually conditioned for operational use, but an estimate of three to four uncrated to one operational unit would not be too far off.

Our team of four civilians, all from Wright Field, was sent to Okinawa and given the responsibility for putting this shipment of VB-3 assemblies in working order and ready for operational use. Our team and work were as follows: the non-technical member, who we called our Coordinator and who handled all the incoming orders and messages, made sure equipment was ready when needed and advised our group of the military action to be taken; the second individual was the bombsight man who took care of the sights and bombing equipment in the aircraft (we had three B-29 bombers); the third, our radio man, had responsibility for the radio receivers in the Razon bombs; and the fourth member of our group was responsible for the gyro assemblies, servo motors, batteries, and other electrical and mechanical parts of the tail assembly including the flare. He also had to make final check of the tail assembly and make sure the bombs were properly installed in the aircraft. It should be recognized that these individuals for the past five years had been project engineers on R&D work on ongoing systems, not necessarily related to controlled bombs like the VB-3. Only one person had worked on the development programs with the Air Force and Gulf Research Company in 1944-45 at Wendover. Hence, it was more of a challenge to use our knowledge rather than experience in making these tail assemblies fit for operational use.

Because these missions used the B-29, and all the developmental work on the VB-3 used the B-17, a new set of trail values for the bombs had to be devised. This took several missions; however, the bombardiers came up with a set of values that seemed to work quite well. This difference in aircraft was a more critical factor for the Razon than it would have been for the Azon bomb.

In September, October, and November 1950, over 30 Razon bombs were dropped on targets in Korea. Most of these targets were rail and road bridges. Where direct hits were made, complete bridge spans were destroyed. Overall accuracy varied from mission to mission but a safe estimate would be when bombs performed properly there was a 50% chance that the bomb would hit the target.

After the mission, when aircraft returned to Okinawa, there were briefings which our group attended. The bombardiers gave an account of the action involved and a general assessment of mission success. Apparently one in seven of the Razon bombs failed to operate satisfactorily. This was just about the number (one in six) that failed during the development phase on the Razon conducted by Gulf Research personnel at Wendover in 1944-45. In short, it seemed as if this failure rate was going to stay with the VB-3. We had a lot of trouble with radios and batteries during this operation and it can be stated that radios gave a considerable amount of trouble during developmental testing. On the other hand, we had no flare failures or pre-ignition of flares.

There was at least one Tarzon bomb dropped during this period and it was against an enemy troop concentration.

A number of recommendations were made by the Bomber Squadron that delivered the Razon. For the most part these concerned redesign of the bomb that would permit more control. Another recommendation was to use B-29s in any future development or tests with the VB-3. As far as one could discern the bombsight attachments worked properly and there was ample time for control as the U.S. had air superiority in Korea. Most of the controllers thought the radios were the weak link in controlling the bombs and they may have been right, we just didn't have any way to tell.

On return to Wright Field our team was asked to brief General Chidlaw and his staff on how the Razon bomb operated in the Korean War. After our third team member gave his report, General Chidlaw said he had heard enough. It was quite obvious he was very unhappy to know the deplorable condition the tail assemblies were in on their arrival in Okinawa and the amount of work required to put them in operating order. Several Directives were issued by

General Chidlaw to correct the situation; however, as happens many times, it was just too late to improve the stored equipment condition very much. We had stated that it appeared only about one-third of the remaining stored assemblies that had not been unpacked could be salvaged.

Perhaps the best action that could have been taken on this shipment of Razons after WWII would have been to have them shipped back to the U.S. and to reopen the Razon Project. The developmental phase could then have been completed and parts, like the radio receivers, could have been upgraded. Post-war thinking changed and very few people thought there would be a need for Razon bombs after 1945, so, as occurs most of the time, the development work never gets completed.

The Bombing Missions and Conclusions on VB-3 Operational Use

Three B-29s were equipped to drop the VB-3. A bombardier was assigned to each plane and he controlled the bomb in both range and azimuth. Most of the bombs were dropped from 15,000-20,000 ft. altitude and, with air superiority, there was adequate time to control the bombs. With regard to bomb results, one bombardier's were very good, one bombardier's were good to fair, and the third, only fair. There was much discussion among these three as to why the differences when the targets were similar and the bombs practically the same. A number of suggestions were made especially with regard to controlling the bomb, but the problem was never completely resolved to everyone's satisfaction.

What the operational use of the VB-3 in the Korean War demonstrated was that there were a number of decisions immediately after WWII that prevented it from attaining its real potential. In particular, the R&D work and testing on the VB-3 should have continued after WWII until it attained a standard level of operational capability. Second, military personnel should have been trained on preparing and maintaining it and to be ready on a stand-by basis for any eventual operational use.

A misunderstanding that occurred was that the VB-3s on hand, i.e., stored in warehouses, were in fair or usable condition. This was not the case as they had not been examined or inspected for use after five years of storage. During this time the conditions under which they were stored were never fully explained, hence we do not know to what temperature or humidity extremes they were exposed.

A recommendation was made to form a squadron of three B-29 bombers with all the required personnel for a 30-day training period. The ground personnel would prepare the VB-3s and the flight crews would use 30 or 40 bombs for actual drop tests. This would take place prior to going into a real-life situation.

The Korean War would have provided a real opportunity for the ground forces and the Air Force to determine the offensive and defensive capabilities of such projects as ground control of the VB-3. Unfortunately, all development work on the VB-3 was canceled at the end of the war, and the opportunity for such a determination never came about.

In short, the abandonment of work on the VB-3 after WWII was not a good decision.

Preparing the Razon for Operational Use

A description of the VB-3 assembly, checkout of components, and related data—part of which is used in the technical manual for operational use—is included here to present in detail the work involved so that one can understand more clearly the overall problems associated with this mission and its success or failure.

The VB-3 Bomb Tail Assembly called the "Razon" weighed approximately 185 lb and was attached to the standard AN-M65 1.000-lb general-purpose bomb and replaced the standard fin assembly. This tail assembly permitted the bomb to be controlled during its flight by means of radio-remote-control equipment located in the carrier aircraft. The bomb flight path could be altered in both range and azimuth by movable elevators and rudders on the rear shroud. Housed in the Razon assembly was the equipment to receive the transmitted radio signals and thus apply up, down, right, or left movement from the electric servo motor to the control surfaces. The VB-3 bombs were equipped with fuses armed with arming wires, both of which were standard ordnance items, and were hung on the aircraft's internal racks as with standard 1000-lb bombs.

During bombing, the bomb was released in the ordinary manner. After release the bombardier became the controller. As he looked through the bombsight he saw an image of the target with another image of the bomb's flare superimposed on or near the target image. This resulted from an attachment called the "CRAB" installed in the bombsight. By seeing both images simultaneously, the bombardier could see where the bomb was going to hit with respect to the target and make the necessary corrections. The flare arming circuit permitted the flare to be ignited 5-10 seconds after release.

The gyroscope unit housed two gyros. The gimbaled directional gyro provided a rotational reference plane for the bomb and maintained this throughout the bomb's flight. The rate gyro provided inputs to the ailerons that controlled the rate of bomb rotation and its position relative to the reference plane, which was approximately ±2.5 deg and at two cycles per second. Figure 10 shows a gyro assembly with cover removed. A check was made with a gyro on a test stand (Figure 11) to determine for the directional gyro that the inner and outer gimbals rotated freely and the proper static balance was maintained. A rapid return to a preferred position indicated a gyro out-of-balance and the adjustable balance weights were moved. When adjustment could not correct the imbalance, another gyro unit was used. This occurred in approximately one in five units. The rate gyro was checked for freedom of motion only.

After the motor switch was turned on, 45 seconds was allowed for the gyros to come up to speed, the directional gyro between 8000 and 9000 rpm (revolutions per minute) and the rate gyro between 6500 and 8000 rpm. The rate limit switches were actuated when the bomb roll rates were between 20 and 25 deg/s. A schematic of the aileron control circuit is shown in Figure 12.

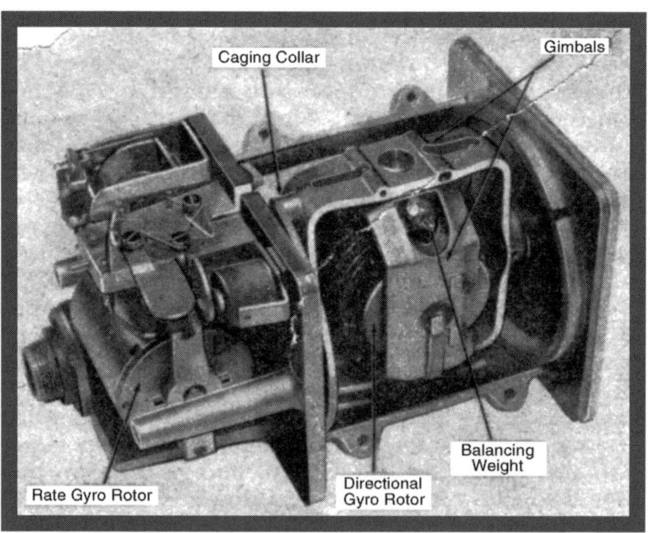

Figure 10. Gyro assembly with cover removed.

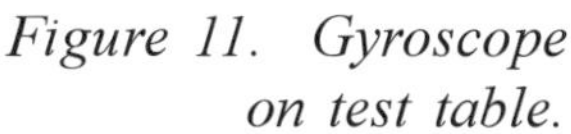
Figure 11. Gyroscope on test table.

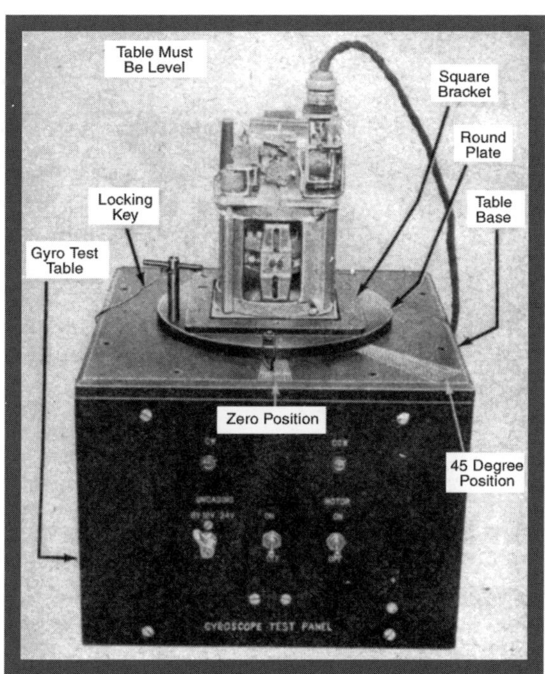

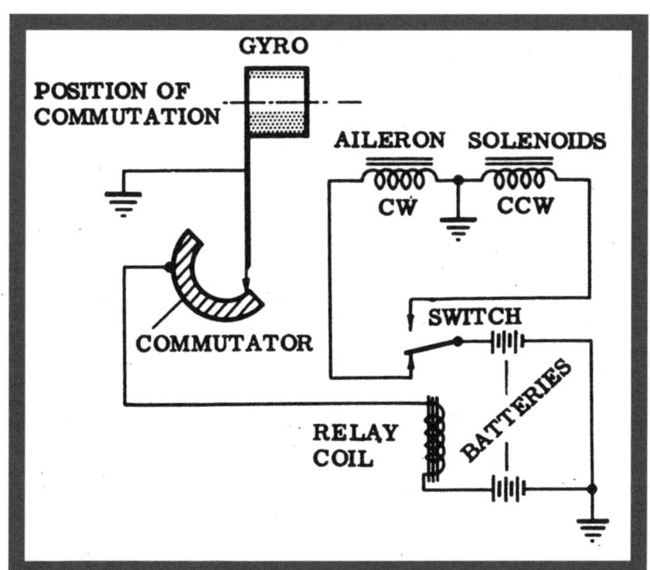

Figure 12. Schematic of the aileron control circuit.

An important check on the directional gyro was a test of the uncaging mechanism, as a prematurely uncaged gyro would mean loss of bomb control and failure to accomplish the mission. The test to determine whether the gyro would uncage under shock was accomplished by placing it on a hardwood surface with the longitudinal axis vertical and, with the top tilted approximately 1 in. from the surface, allowing it to drop. It should not uncage when the test is made from each of the four sides. The second series of tests for uncaging was applying 8 V and 12 V to the uncaging coil. If 8 V uncaged the gyro, a portion of the catch arm was filed, thus moving the plate farther from the solenoid coil and requiring a stronger magnetic pull to uncage. About 10% of the gyros tested required this adjustment.

A check on the rate gyro settings and precession was made while the gyro unit was still on the test stand. As the table was oriented in a position counterclockwise from the zero position, the CCW (counterclockwise) lamp would light, and when oriented in the opposite position the CW (clockwise) lamp would glow. The lamp reversal or equilibrium point would occur within ±0.5 deg from the zero mark and the lamp would flicker rapidly. Changes were made on the adjustment screws on the rate gyro gimbal (see Figure 13). Leaving the gyro uncaged and moving the table at least 30 deg clockwise from the zero mark, the rate gyro was then tilted by pushing down on the side of the gimbal until the indicator lamp changed from CW to CCW. The gimbal should be approximately 3/32 in. above the limiting pedestal. This procedure was repeated, pushing down on opposite sides of the gimbal until the lights changed from CCW to CW.

Calibration was made by positioning the adjustment screws. After making necessary adjustments, a new flicker point was determined which should be 9.5-13.5 deg away from the first flicker point. Next the gyro was re-caged and the gyros turned on, and when they reached operating speed the turntable was set at zero. The voltage was set at 24 V, the gyro uncaged and allowed to spin for 2 minutes, then the turntable was slowly rotated CW and CCW until the flicker point was found. The number of degrees between the position where the unit was uncaged and flicker point at the end of 2 minutes is the amount of precession and should not exceed 5 deg. Most of the gyro units, once they were put into condition, met this precession requirement.

The radio receiving set CRW-7 (Figure 14) in the tail assembly used to receive inputs from the bomb controller operated in conjunction with the aircraft's transmitter RC-186. The receiver operated on a radio frequency determined by a plug-in crystal that set the frequency of the oscillator. The audio-frequency selector filters operated at 475 cps (cycles per second) for left control, at 755 cps for down control, 1900 cps for up control, and 3000 cps for right control.

Power-wise the CRW–7 used 21 V at 3 A, or approximately 60 W. The test equipment (Figure 15) for checking the receiver was made up of a VHF signal generator, two audio oscillators, a milliammeter, a 19-24 VDC power

Figure 13. Top portion of gyro assembly.

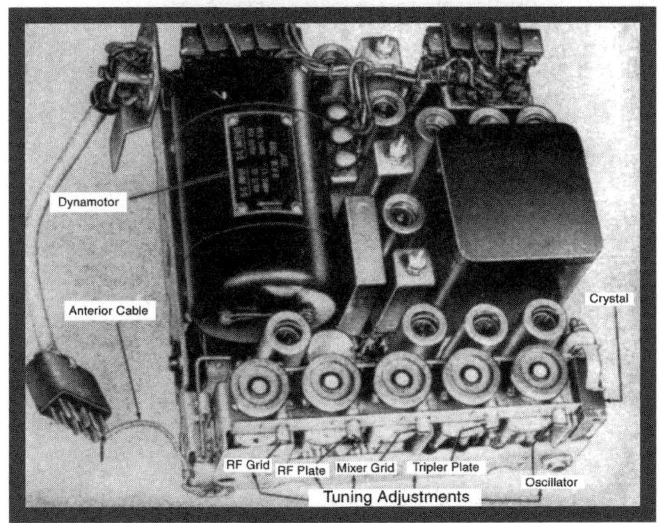

Figure 14. Radio receiver CRW-7, top view.

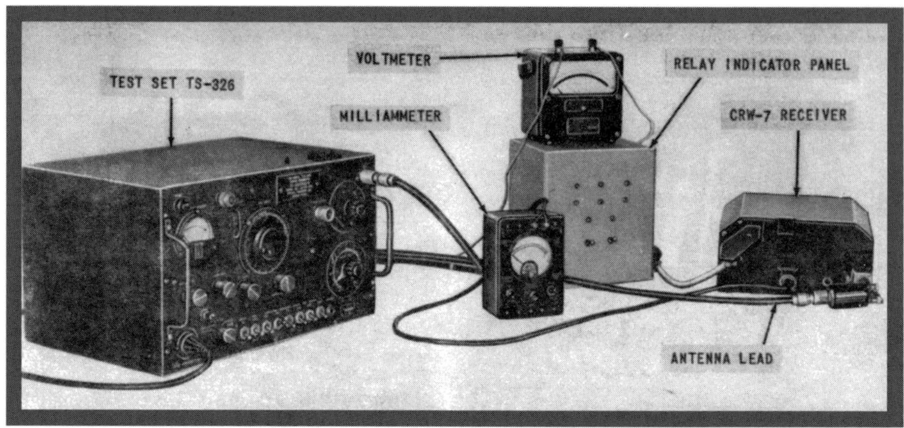

Figure 15. Receiver test equipment.

supply, a headset, and relay indicator panel. Using the relay indicator panel a series of tests were made to ensure that the proper signals operated the relays at the designed levels. Should the relay current be less than the required value, an adjustment of the antenna trimmer control and radio-frequency grid was made. Adjustments were made until maximum output was at least 4.0 mA. About one in four receivers checked failed to operate satisfactorily. This was thought to be due to the rough handling during shipment. One must remember that this equipment used vacuum tube electronics, as all equipment of this type did during WWII, and was quite susceptible to shock. Also, vibration played a part in radio receiver operation, particularly after installation in the aircraft. Of course it was this high number of failures that created the demand for solid-state electronics.

The batteries (Figure 16) furnished power for the tail assembly during flight on the bomb. They were a special miniature type designed to supply 6 A continuous current at 19-24 V for about 5 minutes, which was much longer than required. They deteriorated quite rapidly when stored in a fully charged condition, hence they were filled only a short time before use. Each battery cell consisted of a series of alternate lead and lead-peroxide plates which were separated by a porous material. After the sulphuric-acid electrolyte was added there was a potential difference of 2 V across each cell. It took 27-28 cc of acid per cell which were filled 24 hours before a mission. The batteries had to be placed on charge at a rate of 0.5 A for approximately 16 hours. After charging, the batteries were allowed to cool and then were connected to a load tester (Figure 17). The no-load voltage should be between 23 and 25 V. The load switch was then closed and kept in this position for 8-12 seconds, thus placing a 3 Ω 150 W resistor across the battery and simulating a load current when all components in the tail assembly operated at the same time. The voltage on this test should not drop below 22 V, and if it did the battery should not be used. Sometimes a second charge would bring the voltage up to desired level; if not, it was discarded. About one-fourth of the batteries we tested had to be discarded.

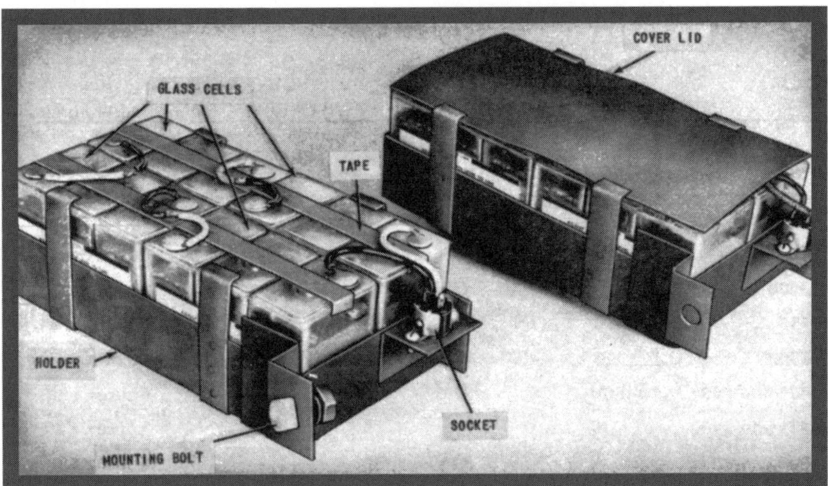

Figure 16. Batteries.

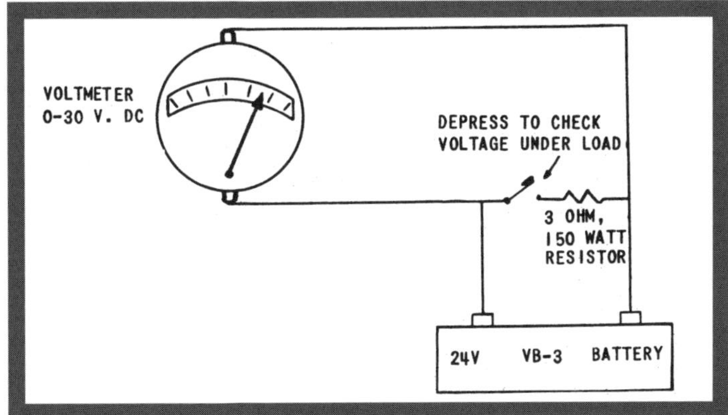

Figure 17. Load voltage test apparatus.

A continuity check (Figure 17A) was made on the flare circuit and this was one item that worked and apparently had a long storage life. We never had to replace a flare.

The servo motors were installed in the tail assembly prior to shipping and through a set of linkages were connected to the control surfaces. Check of the servos was made by using the wiring harness and test panel shown in Figure 18. The assemblies were equipped with cam and rider linkages to limit

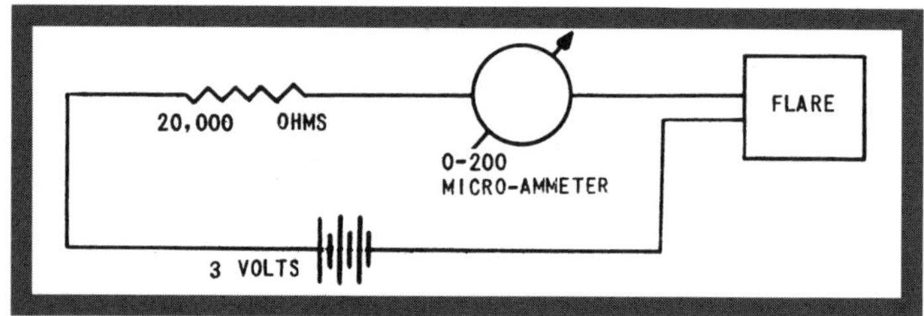

Figure 17A. Flare continuity tester.

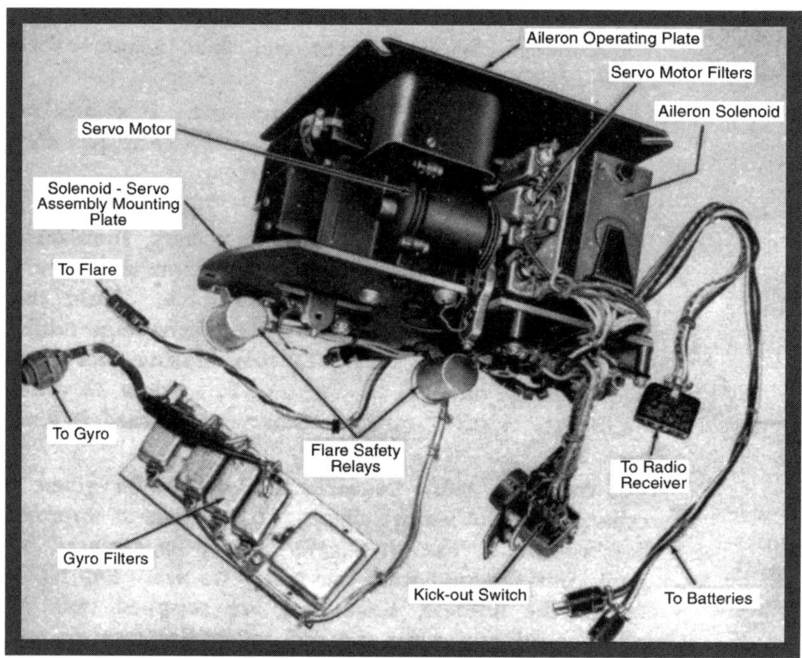

Figure 18. Servo assembly.

the angle of turn of the servo arms. A check of rudder control was made by turning the test panel (Figure 19) switch to the three position, *Left*, *Center*, and *Right*. The two rudders should move together in the same direction. A check of elevator control was made by turning the test panel switch to each of the positions corresponding to *Up*, *Center*, and *Down*. Elevators should move together and in the same direction. A check of the aileron response was made by pressing the test panel aileron switch and observing that the ailerons moved in the proper direction for both clockwise and counterclockwise motion.

On a mission and after takeoff the bombardier turned on a warm-up switch on his panel to warm up the radios and gyros during the flight to target. Also, a crewmember entered the bomb bay, plugged the flare connectors to the flare, and removed the safety pins from the kick-out plug after checking to

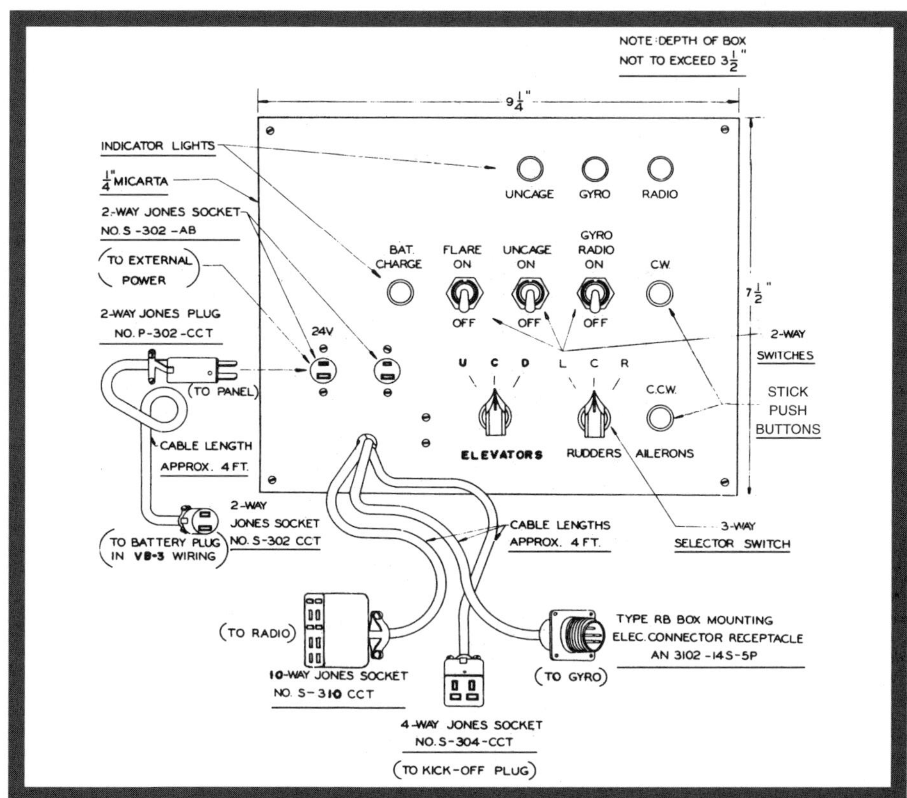

Figure 19. Final test panel.

see that the arming wires were in place. Upon bomb release the kick-out plug was ejected and full voltage was applied to all components in the tail assembly, including uncaging of the directional gyro.

Line of Sight

A sighting method used in the delivery of weapons, known as the "Line of Sight"—a straight line between the eye of the controller and the target—when properly executed, means a hit on the target every time. If the missile or bomb is controlled in some manner so as to stay on this line from the point of release until it impacts at the end point, then it is obvious it will hit the target. In a sense, the line of sight could be called a system of "can't miss." Perhaps the greatest reason this technique was not used was the difficulty of getting and keeping the controlled body to stay on the line of sight (see Figure 20).

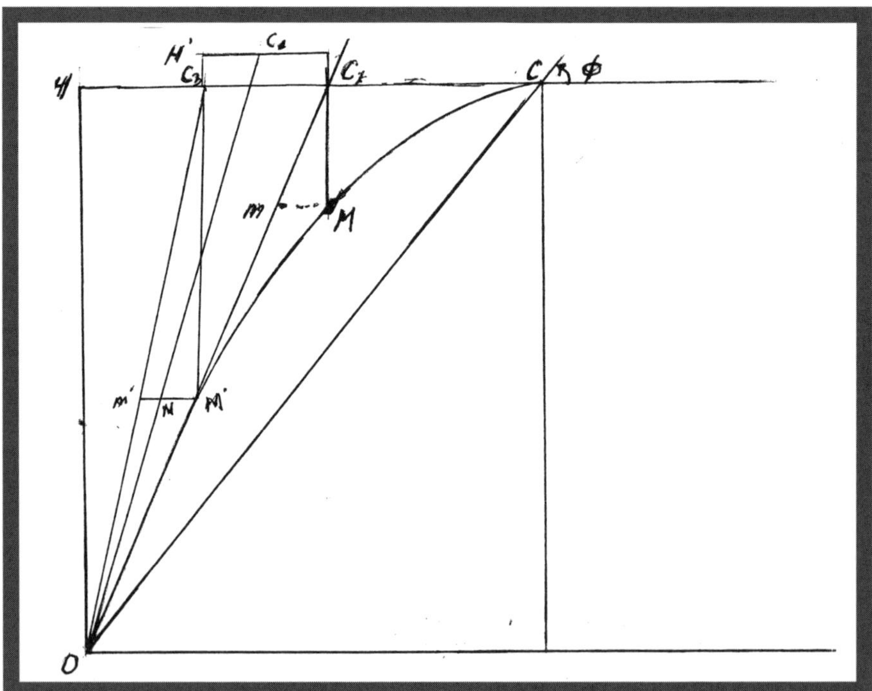

Figure 20. Line of sight diagram.

In the Gulf[2] Progress Report for Testing the VB-3 Razon Bomb from April to July 1945, the following statement was made: "If with larger fin surfaces or other means a 100% higher bomb maneuverability could be obtained, then there is a possibility of controlling the bomb by a line of sight method. Since this system involves no requirement or pre-knowledge of the time of flight, higher range accuracy may be obtained with it. As a long-range post-war project, it would seem, therefore that this general line of approach might be worth investigating carefully."

At this point it should be stated that during WWII the Gulf Research and Development Company was under contract with the National Defense Research Council and submitted Progress Reports every three to four months; however, all the work and reports were classified and it wasn't until years later that these reports were unclassified.

Consequently, because of the classification and limited distribution of Gulf's reports, many in the military were unaware of the recommendations that had been made. However, with the event of the Korean War in 1950 there was a renewed interest in guided bombs and the Controlled Aircraft Branch at Wright Field undertook a project to study the feasibility of applying line of sight to vertical bombs, particularly the VB-3.

Our unit had two aircraft assigned to it for test work—a B-17 and a B-29. After discussing with our pilots the line of sight problem and the various aircraft maneuvers that might be involved it was decided to use the B-17.

Our first objective on this project was to obtain data from flights and make calculations on and for the various design parameters; for example, how far from the line of sight would the missile be at any time and what would be the magnitude of control to put a bomb in the proper position. Thus we had a three-coordinate system and it seemed the first that should be determined was the aircraft. We would make ten runs at altitudes of 15,000 and 20,000 ft. From a review of preliminary data we recognized that the bomb would naturally lag the aircraft; however, if the aircraft airspeed could be reduced after bomb release it would aid in providing a part of the overall solution.

Our data collection was by personal observation, which for this phase was quite adequate. The flight plan was to make a standard bomb run (bomb-bay doors open) and at bomb release the pilot would pull back on the control column and at the same time the copilot would reduce engine power to reduce airspeed. Thus there would be a gain in altitude and a reduction in airspeed,

[2] Gulf Research and Development Company

all this to take place in 10 seconds at constant heading after which the pilot would return to level flight and the new airspeed. In this time we planned to have the technician read the instruments. The pilot was to announce when 10 seconds had passed.

At each altitude we made ten runs from which the data point values could be averaged. A plot is shown in Figure 21 which, in terms of flight path, is practically the same for the two altitudes.

On the first set of calculations for aircraft position the method of least squares was used; however, it seemed more practical to use a point-by-point value especially in light of the aircraft changing airspeed. On the 20,000-ft. altitude plot, 40 increments were used; however, Figure 21 gives only the first ten after bomb release.

As with any analysis it is best to start with the simplest and build up to the more complicated. In the line of sight problem it was apparent that each part had limitations and these could be expressed as parameters. Then we must determine their limitations and what changes could be made which would meet the basic line of sight requirements. For example, the target would be fixed on the ground, hence the position of the aircraft and bomb must be varied with time and be colinear to be on the line of sight.

A no-wind condition was assumed in all calculations and hence no drift in the bombing aircraft. Conservative values for bomb trail and time of fall kept things simple.

In order to establish an early reference we considered a normal bomb drop from approximately 20,000 ft. and made calculations based on its most probable trajectory shown in Figure 20; therefore only plane of action is required.

The two-plane concept

The first plane would release the bomb in the normal manner. The second plane flying directly behind and slightly higher than the first plane would control the bomb. Since the bomb would be ahead of the line of sight of the second plane it could easily be controlled to the desired position on the line of sight. The lead plane could have dropped two bombs and it would have been a fairly simple matter to put two controllers in the second plane, thus two targets could have been bombed instead of just one. Of course there were other advantages, e.g., if one bomb failed, the second could be used as a backup; also, multiple hits on a target would be more likely to do the most destruction.

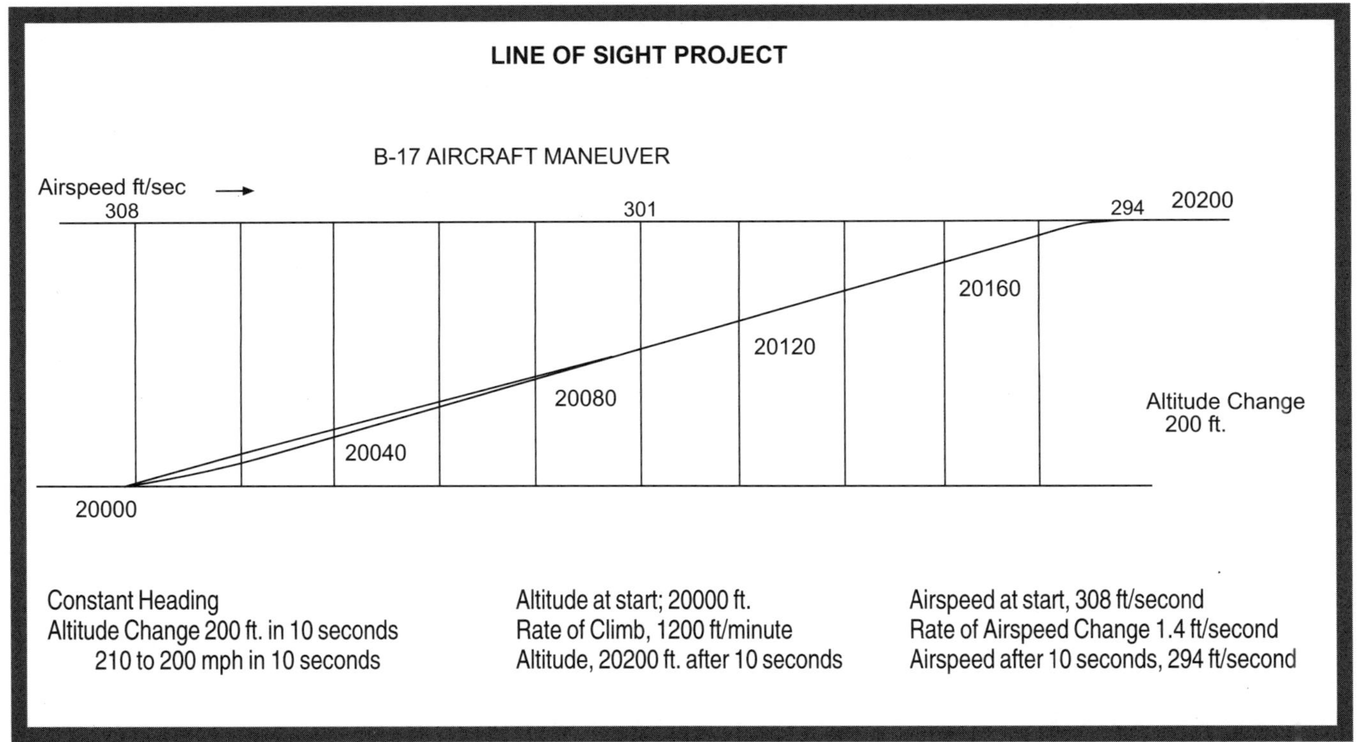

Figure 21. Line of sight project.

This concept was only discussed and to my knowledge nobody ever followed up on it. However, years later the AGM-62 Walleye Missile was developed and used in the Vietnam Conflict with the pilot in a second trailing aircraft controlling it.

Line of Sight Equations

This is a three-coordinate system, with aircraft coordinates being a_x, a_y, bomb coordinates x_b, y_b, and target x_t, y_t; the target x_t, y_t is at the origin, or 0,0. The aircraft and bomb coordinates vary with time; however, the altitude a_y is fixed. The aircraft velocity is fixed at its true airspeed, in this example 210 mph. The range is

(time of fall) x (aircraft true airspeed V_a) or 35 x 210 x 1.47
308 x 35 = 10,780 ft.
Trail = 1400 ft.

During this time the bomb travels 10,780 – Trail = 9,380 ft.
The aircraft coordinates for any time t after release is Range – $V_a t$
Hence, $x_a = 10{,}780 - 308t$, $y_a = 19{,}723$

The bomb coordinates at release are the same as the aircraft; however, after release the forward velocity is $308t$ – Trail at time t for each second the bomb trail increases 40 ft.; hence this is $40t$ or the horizontal component

$x_b = 9229 - (308 - 40)t$
$x_b = 10{,}780 - 268t$

If the bomb drops without resistance, then its travel after release is $\frac{1}{2}gt^2$ and its coordinate $y_b = 19{,}723 - \frac{1}{2}gt^2$ where g = 32.2 ft./sec^2.

Hence for our three-coordinate system we have:

$x_a, y_a = 10{,}784 - 308t,\ 19{,}723$
$x_b, y_b = 10{,}780 - 268t,\ 19{,}723 - \frac{1}{2}gt^2$
$x_t, y_t = 0, 0$

Three points are colinear if the value of the determinant = 0

or
$$\begin{vmatrix} x_a & y_a & 1 \\ x_b & y_b & 1 \\ x_t & y_t & 1 \end{vmatrix} = 0$$

since $(x_t\, y_t)$ is (0, 0)

$x_a, y_b - x_b, y_a$: and $x_b = x_a\, y_b \,/\, y_a$
or $x_a = x_b\, y_a \,/\, y_b$

A number of calculations were made to determine the unknowns x_b and y_b in terms of known values for both the maneuvering and non-maneuvering aircraft. Using Figure 20 for the non-maneuvering aircraft and after 25 seconds, it would be at C_3, and OC_3, the line of sight. At this time the missile is at M′ and the horizontal distance is m′ M′ from the line of sight OC_3. If the maneuvering aircraft gains altitude to H′ and slows down it would be at position C_4 and the line of sight would be OC_4. Thus NM′ is the required distance for the bomb to be in the line of sight and is approximately 2/3 the m′ M′ value. This would be in line with the calculations made for the maneuvering aircraft and bomb position after 20 seconds; magnitude-wise this would be 400-600 ft. for NM′ using the above flight conditions.

Line of Sight Diagram

In order to simplify this problem and explanation, a number of assumptions were made, namely a vacuum atmosphere, two dimensions, and aircraft speed (velocity) and altitude that remain constant after release of the missile.

Figure 20 illustrates geometrically the line of sight and related factors involved. If we use this diagram for our coordinate system we can say OH is the altitude, equal to A; let C represent the position of the aircraft at missile release point, hence HC is the range R, and let O represent the target. M represents the missile which follows or moves along a parabolic trajectory. The distance $C_2M = \frac{1}{2}gt^2$ and Y coordinate is $A - \frac{1}{2}gt^2$, X coordinate is CC_2 or Vt. The aircraft position after time t has moved to position C_2 and the distance covered by aircraft flying with velocity V is $CC_2 = Vt$. Since range is R then X coordinate would be (R–Vt) and Y coordinate is altitude A or (R–Vt, A). The missile's desired position after time t to be on the line of sight is m. mM is the distance necessary to add to C_2C for our solution, hence missile coordinates should be $(C_2C + mM,\ A - \frac{1}{2}gt^2)$. Let the target be at origin whose coordinates are (0,0). Substituting coordinates we can find x_a and x_b that will satisfy the line of sight requirements.

A series of calculations were made using these coordinates and data from the flights involving a reduced airspeed after bomb release. Based on the results of these calculations, a number of conclusions were made, as follows: Slowing the aircraft probably provided the most promise; however, calculations showed it would have to be reduced to approximately 185 mph and hopefully in 20 seconds to allow control for the next 15 seconds; control of the bomb for the values we had on the past tests of the VB-3 provided only about a third of the control that would be required; it might be possible to design a new lift shroud but more tests, especially wind tunnel, would be needed.

The final conclusion was that aircraft maneuver after bomb release combined with a 25% increase in bomb control would permit line of sight control time period of 15-20 seconds, which would be the time period before impact. This would probably be sufficient to hit most targets. Another proposal was to move the controller behind the bomb bay, i.e., control the bomb from the rear of the aircraft. Many suggestions were made to improve the above technique and undoubtedly these would have provided partial solutions; however, the project was closed out because of higher-priority work.

The Bomb Flare

A pyrotechnic flare was used on a number of vertical bombs as a means of tracking the bomb after it was released. The flare had a burning time of approximately 1 minute. The Azon, Razon, and Tarzon bombs used this type of flare. Most of the flares used gave off a white light. Some red and green flares were used for certain experiments and on special occasions. Further, only white flares were used on the controlled bombs in the Korean War. The flare was an item of considerable concern because of the possibility of pre-ignition, as happened on one flight when the control bomb never got out of the bomb bay but the flare came on. The pilot thought the airplane would catch fire and had the crew bail out. However, the crew chief tried to battle the flaming flare and eventually the pilot and the crew chief landed the airplane. What happened was that the dummy bomb failed to release and the control bomb which was above the dummy released and after 5-6 seconds the flare ignited. The dummy bomb got the control bomb wedged in and it could not fall properly. The difficulty and precautions taken with flares is not written up in the Gulf Progress Reports. It should also be mentioned that during ground tests of flares we had one ignite and this was thought to be due to an electrostatic charge.

In the Razon Maintenance and Checkout Manual a note of caution is given as follows:

> Do not use the common ohmmeter method of checking continuity. The maximum safe current for the flare is only one-third of a milliamp (approximately 1/3000 of an ampere). The output of common ohmmeters generally exceeds this limit. An exploded flare will result.

A diagram of the continuity check is shown in Figure 17A.

In 1949 the Control Aircraft Branch in our Laboratory obtained data on a recently developed high-intensity electric lamp. Based on this data it was decided to set up a project to determine the feasibility of using an electric lamp as a substitute for the bomb flare. Three lamps were procured and tested. The cone of intensity was approximately ±5 deg and it was clearly visible at 20,000 ft. A contractor was given the task of devising a mechanism to keep the light pointed in the direction of the aircraft because of the lamp's narrow beam and the change in horizontal angle as the bomb moved through its trajectory.

Another problem existed: As full control would be applied on a controlled bomb, the bomb's angle of attack would change about 10 deg; consequently unless the mechanism was synchronized to change when control was being applied, the lamp would not be visible. In an effort to keep the costs down it was decided the mechanism to be used would move only an amount to make up the change in trajectory. Two bombs were equipped with this mechanism and tested at Holloman Air Force Base, Alamogordo, New Mexico. Test results showed the flare to be visible for about 25 of the 35-36 seconds of flight. In addition to the mechanism for moving the flare in flight, a battery power supply was used, as the electric lamp for commercial use required 110 V. These tests proved that an electric lamp would have the intensity but there were many other considerations such as an improved mechanism to change lamp angles at various times, and the added complexity seemed to create more problems; hence no further efforts were undertaken.

The Spazon Bomb

The purpose of this project was to determine the amount of dispersion between controlled bombs when released at the same time. Hence, the bombs assigned to this project were called Spazon bombs.

Although there were two efforts on Spazon bombs, one by the Gulf Research and Development Company working under an NDRC contract, and called the Gulf approach, and one by the Military, they varied considerably in content. Gulf derived some techniques but never fully implemented or tested these on real bombs. Further, the work Gulf had planned used the Razon bomb, whereas the Military used the Azon.

Gulf did not specifically describe the techniques they would use; apparently from some of the test work on Razon bombs they would spin the bombs after release for a time, then apply control. This spinning would tend to keep the bombs closer together immediately after release, then it was supposed they would use cluster control, i.e., control the mean value of control on the bombs after the spinning action. If this were the approach, only tests could determine how well their techniques could be applied. A number of statements from Gulf's Progress Report on Razon bombs are included here to indicate the possible approaches that could be taken on Spazon bombs.

> "From each of the three altitudes of 10,000, 15,000, and 20,000 ft., two Razon and at least four standard bombs were released for the purpose of obtaining the time of fall and trail of the Razon bomb when uncontrolled.... In order to eliminate effects of dispersion, the Razon bombs were made to spin at a controlled rate of 90° per second throughout their entire flight.... In the next series of tests the bomb was made to spin for the first 15 seconds of flight and then erect before the control was applied, using the techniques developed for the Spazon bomb."

Undoubtedly Gulf could have provided the technique for using the Spazon bombs; however, higher-priority work was probably the reason this work was not carried on.

The advantages of the Spazon bomb are worth noting before describing the Military part on this effort. The reliability of a controlled bomb to operate as it should 100% of the time is asking a good deal more than was shown during the developmental stage. In other words, one in five or six bombs failed in some measure to operate as desired. Therefore if two bombs were dropped, the probability of at least one functioning properly would be much greater. Also, if enemy fire prevented more than one pass at the target, two bombs would be more likely to score a hit than just one.

In the Military project on the Spazon bomb, three Azon bombs were used and dropped simultaneously from a B-17 aircraft from an altitude of 20,000 ft. and an indicated airspeed of 180 mph. Three different colored flares were used, namely white, red, and green. It was decided to use a second controller (in addition to the bombardier). This controller had two control boxes and sat next to the bombardier. He controlled the bomb with the white flare. The bombs were dropped in minimum train, i.e., the interval between releases was a minimum. The purpose of this was that bombs dropped in salvo would sometimes hit another bomb on release or just on leaving the bomb bay. The dummy bomb was the first to be released, then the three Azon bombs. Obviously the dispersion of Azon bombs would be different for Razon bombs but it was thought these would provide good data.

The bombs were dropped as planned but about 10 to 15 seconds after release the bomb with the green flare took erratic action. It appeared as if the controls were frozen as the bomb moved off course without any control signals being applied. The other bombs responded to control and continued on a stable trajectory. After impact the Bomb Range crew reported that the bombs with the red and white flares were within 50-60 ft. of each other in azimuth and 200 to 300 ft. beyond the target. The dummy bomb hit about 500 ft. to the right of the target and about 25 ft. short. The bomb with the green flare went completely out of control and hence its impact point would be of little or no value.

Based on these results it was concluded that the dispersion of a multiple release of two controlled bombs would not be very great, particularly in azimuth. However, if dual controls were applied, as would occur with Razon bombs, the dispersion could increase. Therefore it was recommended if the Spazon project were to continue that tests be conducted with the Razon bomb, and also that other techniques such as those developed by Gulf be evaluated for possible use. The work by the Military on the Spazon project was last to be performed and the project was closed out.

VB-6 Felix Guided Bomb

Although the Felix is referred to as a guided bomb, it is more of a target-seeker than a bomb, as it depends on outside signals to guide it to the target. It probably should have been called a target-homing bomb, since it didn't need or require an external guidance input. The infrared or heat sensor was located in the nose section and in the early models was a four-quadrant arrangement, i.e., it responded to up-down and right-left. In laboratory models the nose would follow a heat source as it moved about, but if the source were moved a fair distance, the sensor had difficulty following it.

Figure 22. VB-6 Felix guided bomb.

The Felix bomb development never progressed to the same level as some of the other vertical bombs but, as happened in other programs, the technology used supplied some of the building blocks to be applied on other programs—in this case it was the Falcon missile built by Hughes Aircraft Co.

Designers of the Felix bomb thought the greatest possibility for use of a heat-seeking missile would be against factories having considerable heat output and also against steel mills. The VB-6 project was closed out at the end of WWII.

VB-9 Vertical Bomb

The VB-9 was in a sense a forerunner of the MACE missile. The original version of the VB-9 relayed a radar picture of the target area to a controller or bombardier who presumably knew how the target would show up on his viewing screen. He would then send radio control inputs, or signals, to the bomb to guide it into the target. The nickname ROC for the VB-9 was taken from mythology—a giant bird so large it could carry off elephants. The wings, nose, and tail sections were attached to a 1000-lb general-purpose bomb. In late models the four wings were replaced by circular airfoils. Also other homing or seeking devices were being considered for this VB-9. Development work was never completed and the project closed out at the end of

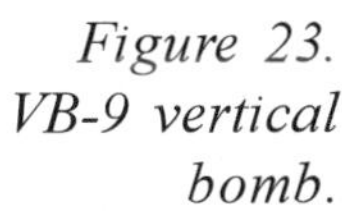

Figure 23. VB-9 vertical bomb.

WWII. The main difference between the VB-9 and the VB-6 was that the VB-9 had such large control surfaces that it might be considered a glide bomb. Further, it sent a picture of possible targets for the bombardier who in turn could alter its trajectory by radio control.

VB-10 Television Bomb

The VB-10 as a television bomb had the potential to become one of our best WWII weapons. By the same criteria early development work also showed that it would probably be the most difficult to develop. Once developed, however, it could have become our "Pseudo Kamikaze," as the guidance system took the bomb right to the target. The technology that this and similar projects have provided has produced the "Smart Bombs" currently in use. As a television radio-guided bomb, the VB-10 had many advantages over other types of controlled bombs. The first and foremost feature was that once the controller had identified and locked-in on the target, he could control the bomb until it hit the target. This assumed that everything connected with controlling the bomb worked properly. Initially the VB-10 was envisioned as a television bomb with a camera in the nose that would relay a picture to a controller who in turn would guide the bomb to the target. Having a sighting mechanism that let you dive on the target gives rise to the term "Pseudo Kamikaze." Design techniques of using television guidance were later incorporated in the Rascal Missile built by Bell Aircraft.

Shown in the photograph of the VB-10 (Figure 24) is a circular airfoil mounted at the bomb's center of gravity. This airfoil, approximately 5 ft. in diameter and 30 in. in width, provided the up-down and right-left control of the bomb. Control rods electrically driven moved the airfoil in relation to its position on the bomb thus providing lift or whatever control was demanded.

The VB-10, with a tail assembly that was intended to keep the bomb stabilized in roll, was a little over 10 ft. in length. The bomb was a 1000-lb high explosive. Although the ideas sound simple as to bomb control, the design and

Figure 24. VB-10 television bomb.

implementation of equipment required of a television bomb becomes quite complicated and a number of these were not incorporated.

At launch, the VB-10 was in a horizontal position and at impact its longitudinal axis was near 70 deg to the vertical, hence it must change 70 deg. If the camera was mounted on a stable platform in the nose and aimed directly forward it would have had to be slaved 70 deg during the flight time of the bomb. Also, to keep the camera stable required that it be mounted on a stable platform whose rotational motions were not coupled with those of the bomb proper. A directional system similar to that on the Razon was also required in order to have the proper orientation for up-down and right-left control. Another most desirable arrangement which was used on the Rascal missile was control of the television antenna in the direction of the aircraft with the controller.

The VB-10 had several advantages over other types of controlled bombs that are worth noting: (a) because of the magnitude of control that the VB-10 had, the release point would not be as critical as with other bombs, and (b) it was one of the few types of radio-controlled bombs that did not require the aircraft to remain on steady conditions after release; i.e., the aircraft after release was free to take evasive action or make maneuvers that would minimize enemy fire, particularly from the ground.

During 1944-45 two VB-10s were tested at Wendover. A remark by one of the controllers was that it flew all over the sky. Obviously much development would be required to make this bomb operational. The project was closed out in 1945.

VB-13 Tarzon Bomb

The VB-13 was essentially a British 12,000-lb Tall Boy bomb with a tail section having components similar to those in the Razon bomb. Mounted near the center of gravity was a circular shroud to provide added lift when an increase in pitch (range) was desired. The tail assembly and lift shroud weighed almost 1100 lb, making the total weight approximately 13,000 lb, thus the name VB-13. It was radio-controlled by the bombardier in range and azimuth and, as with other controlled bombs, had gyros for reference and rate of rotational motion. Aerodynamic control was achieved by movable surfaces on the tail assembly that also had ailerons for roll control. Unlike the Razon, which had electric motors to move the surfaces, the Tarzon used a pneumatic system of compressed air. In the early development stages the air servo motors would occasionally freeze up. This was eventually overcome by using a dry nitrogen supply.

Figure 25. VB-13 Tarzon bomb.

A pyrotechnic flare attached to the tail assembly permitted the bombardier to track the bomb in flight.

The Bell Aircraft Co., Buffalo, New York, had the contract to furnish the control assemblies attached to the Tarzon.

Several bomb tests were made at Alamagordo, New Mexico, in 1950 using a B-29 aircraft. Although the Tarzon had been deployed to Okinawa for possible use in Korea, only one or two were actually dropped. The Tarzon had not been fully developed or tested when the project was terminated in 1951.

Foreign Bombs

German Fritz X

The first series of Fritz X were produced in February 1942 and were followed by the prototype series in December 1943. Development and testing for tactical employment was in the summer of 1943, although mass production was not reached until January 1944.

The two photos show the Fritz X and installation on carrier aircraft which could have been a Dornier 217, Heinkel 177, Junkers 290, or a Junkers 188.

This bomb was designed for use against armored naval and land targets and was the first guided missile to be used operationally by the Germans. It was

Figure 26. Fritz X.

Figure 27. Fritz X on carrier aircraft.

never extensively used because of the small number of aircraft available with which it could be used. Accuracy was not high, averaging about 20% against shipping. This leads one to believe the bombing technique and release by a bombsight was similar in design to those used by the Allies, but that the sighting method for controlling the bomb to target wasn't too effective. Radio control was used initially, then in later versions they switched to wire control. We had a captured tail assembly for study and evaluation at Wright Field during WWII. The controls for guiding the bomb were similar to those used on our Razon bomb; i.e., they were electrically operated and installed on the tailing edge of the octagonal tail assembly. The inner controls of ailerons on the rear assembly were apparently used to control roll. On the extreme rear of the tail assembly was a unit that looked as if it had an electrical lamp assembly attached; if so this could have been used during bomb control of the trajectory.

Some general characteristics were as follows:

Length: 10.4 ft.
Span: 1.8 ft.
Total weight: 3450 lb
Weight of bomb: 2530 lb
Tail weight: 260 lb
Weight of explosive: 660 lb
Fuse: Contained a self-destruct fuse to destroy the remote control section
Speed: Estimate from an altitude drop at 13,100 ft. was 820 ft/sec
Estimate from an altitude drop at 26,200 ft. was 950 ft/sec

Later models were fabricated but not used operationally. The X-2 and X-3 resembled the Fritz-X. The X-3 had two wings and X-6 a high explosive charge.

There is no known data on how effective this bomb was in operational use, nor its overall accuracy.

German Hs293D

This was a radio-controlled bomb with a television pickup installation and was intended to be used against land and non-armored naval vessels. The television units were still in the development stage at the end of WWII.

Figures 28 and 29 show the main features of this bomb. It has been included here to show that in general all the countries involved in producing weapons of war were exploring the possibility or feasibility of using television technology in the design of their missiles.

Figure 28. Two Hs293 A-1 bombs on carrier aircraft.

Figure 29. Hs293 A-1 bomb.

General characteristics of the Hs293D:

Overall length: 16.6 ft.
Span: 10.2 ft.
Total weight: 2390 lb
Bomb load: 1335 lb
Explosive weight: 650 lb
Engine weight: 295 lb
Engine performance: 14,350 lb/sec
Initial speed: same as carrier aircraft
Maximum speed: 790 ft/sec
Carrier aircraft: Dornier 217 and Heinkel III
Release altitude: 985 to 19,700 ft.
Range distance from release: 2.5 to 8.7 miles

Note: Using an engine permitted the missile to move ahead of the aircraft, thus giving the controller the ability to use the line of sight technique for steering the bomb to the target.

Japanese Kamikaze

The Japanese Kamikazes have been included in this report because the Japanese modified their fighter aircraft to perform in the same manner as we would a guided vertical bomb. Thus the aircraft had duel functions as a dive-bomber and as a bomb itself on a one-way suicidal trip. The word "Kamikaze" had its origin in early Japanese history meaning "Divine Wind."

Based on American naval records it was not difficult to see why Admiral Nimitz and others were quite concerned about the Kamikaze threat. More American warships had been sunk or damaged in three months during the Philippine and Okinawa operations than had been sunk and damaged in all previous naval engagements of the Pacific War including Pearl Harbor. The Kamikaze had to be taken seriously. Perhaps its effectiveness as demonstrated in WWII will lead to some thought in the future of a robot-type Kamikaze.

The closest we had to the Japanese Kamikaze was the VB-10 vertical bomb, later dubbed the "Pseudo Kamikaze." Once the television technology was available and applied to the bomb guidance system it became a "smart bomb," and what was started some 50 years earlier was used quite successfully in the Gulf War.

Figures 30, 31, and 32 show some of the action with and against Kamikazes. It was a stripped-down version of the Mitsubishi A6M Zero-Sun and called by some the Zero fighter.

Several general characteristics were:

Type: single-seat interceptor fighter/light bomber

Performance: maximum speed 346 mph at 19,700 ft.; maximum range 1120 miles

Weight empty: 4178 lb

Dimensions: Span: 36 ft.
Length: 29 ft.
Height: 11 ft. 6 in.
Wing area: 230 ft

Figure 30. Anti-aircraft fire shot down suicide plane.

Figure 31. Kamikaze pilot tries to maneuver his "Zeke."

Figure 32. "Judy" pilot to flaming death.

Chapter 3

Post-WWII Guided Missiles

The Matador Design, Test, and Deployment

It was some time before the builders of guided missiles realized that autopilot components could not be substituted for similar components in a missile flight control system. It took a number of flight failures before the obvious got to the forefront. The autopilot came about as a need to have a stable platform for bombing; on the bomb run just prior to release of the bomb, the same set of equipment could provide pilot relief for short periods, as long as flight conditions didn't vary greatly. The missile flight control system had to produce missile stability over a wide range of flight conditions and had to be reliable. Hence reliability became a design requirement rather than just another desirable characteristic. In most cases the reliability needed could be attained only through redundancy. This was a whole new ball game and now the event of flight control system design became an input to whether the missile flight would be a success or failure.

Much has been said and written regarding the propulsion and guidance systems in missiles, but not too much on the system that provided the stability and control required to move along a given flight path. The Matador Missile was no different than others in the early stages. Originally when autopilot components were used they worked up to a limit, but when that limit was exceeded they failed. A case in point—the vertical gyro in the early Matador was a Pioneer Gyro from an autopilot system; it wasn't designed for a missile system and it failed right after launch.

A second item was the electrohydraulic control valve. This valve replaced the solenoid-type G-1 autopilot valve which was used on early Matadors. The replacement valve was a two-stage electrohydraulic control valve which was a new design that used hydraulic amplification to attain the power gain between a low voltage input and a large force output. This type of valve was eventually used on a number of missiles and flight control systems. The two-stage valve not only had the performance but it demonstrated an increase in reliability.

The Matador was a high-"T"-tail-type aerodynamic vehicle. It didn't require maneuvering of any magnitude after launch and had the entire tail move for pitch control. Yaw and roll control was by a very simple arrangement of spoilers, one set on each wing. One actuator was used for both spoilers as it moved in one direction to displace one set of spoilers and in the opposite direction to move the other set.

The Matador was subsonic and was powered after launch by a J33 engine. Range was 300-500 miles and it flew at an altitude of approximately 30,000 ft. A number of guidance systems had been proposed but were not fully developed, and radio control was used. Early models of the Matador were launched at Alamagordo, New Mexico, in 1949. As mentioned in another portion of this book, the designers thought aircraft autopilots would furnish the desired requirements for flight control; however, they miscalculated terribly in that those items had neither the performance nor the environmental characteristics needed for a guided missile like the Matador.

The first two Matador launchings ended in crashes about 500-600 yds from the launcher. The vertical reference had been lost right after launch, which in turn meant the vertical gyro failed. Thus followed a suggestion to the contractor by Air Force personnel monitoring the project to use a certain gyro they had tested and which would meet the launch and flight requirements. In essence this centered around the uncaging mechanism that was not activated until time zero at launch. This particular vertical gyro was adopted by the contractor and used on all the missiles they delivered for use, both training and operational. Further, this gyro was hermetically sealed and not subject to dirt, dust, or moisture which could foul up the bearings or other parts of the gyro.

A schematic of the flight control system is shown in Figure 35. Details of the electronic equipment shelf are not given in this diagram; however, the amplifiers for roll, pitch, and yaw were mounted on individual chassis. Also included on this shelf were the computational circuits and electronics used for guidance, airspeed, and altitude controls. All the electronic chassis were mounted on vibration isolators. Despite many redesigns and changes in equipment, most of the equipment on this shelf failed to pass vibration tests, especially after third or fourth start-ups. Further tube selection was resorted to as the only method to obtain the performance required in one set.

There were so many failures on the early Matador launchings that the contractor personnel facetiously started to call it the "El Clobberdor." However, after replacing the vertical gyro and several other flight control components with more reliable ones, the next series of launchings were all successful.

Design Characteristics

Span: approx. 28 ft.
Length: approx. 40 ft.
Height: approx. 9.5 ft.
Weight: approx. 13,500 lb
Speed: approx. 600 mph

Right after launch the TM-61A Matador went into a climb to reach its cruising altitude of 30,000-40,000 ft.

In March 1954 the Matador was deployed in Europe and supported the NATO forces in West Germany. In 1955 a number of Matadors were launched on the test range in Libya. As part of a test and training program, Matadors were assembled and checked out at Wheelus Air Force Base near Tripoli and trailer transported to a launch site in the desert. The final flight phase, i.e., a dive on target from cruise altitude, was not too successful as there was quite a bit of scatter between missile impacts and some missiles even broke up before impact which indicated a flight flutter problem. Finally, as a solution the wings were blown off right after the dive signal was given. This proved to be more reliable and the warheads followed a ballistic trajectory to a predictable impact point.

TM-61 missiles were also sent to Korea and Taiwan. In 1957 the Glenn L. Martin Co. delivered the one-thousandth Matador to the Air Force. In 1959 a phase-out of the Matador was started and completed several years later.

Matador MACE

Technology developed for the TM-61A was transferred to the design of the Matador MACE Missile, Figure 34. The MACE, like its predecessor the Matador, was a tactical surface-to-surface-launched missile designed to destroy ground targets. It was first designated the TM-76 and later the MGM-13. It was launched from a mobile trailer or from a bomb-proof shelter by a solid-fuel rocket booster that dropped away after launch. A J33 jet engine powered the missile to the target. Development of the MACE began in 1954 and the first launch was in 1956.

The MACE was developed in two versions, the "A" and the "B." The "A" employed a terrain map-matching radar guidance system known as ATRAN (automatic terrain recognition and navigation) in which the return from a radar scanning antenna was matched with a series of radar terrain maps carried on board the missile that corrected the missile flight path if it deviated from the map. The "B" version used a jam-proof inertial guidance system and had a range twice that of the "A" version.

MACE "A" missiles were first deployed to USAF forces in Europe in 1959. These remained in service until the mid-1960s when some were used as target drones because their size and performance characteristics resembled those of a manned aircraft. Development of the "B" missiles began in 1964 and these remained operational in Europe and the Pacific until the early 1970s. The MACE "B" shown in Figure 34 was based on Okinawa before coming to the Air Force Museum.

Figure 33. Matador TM-61A.

Figure 34. Matador MACE.

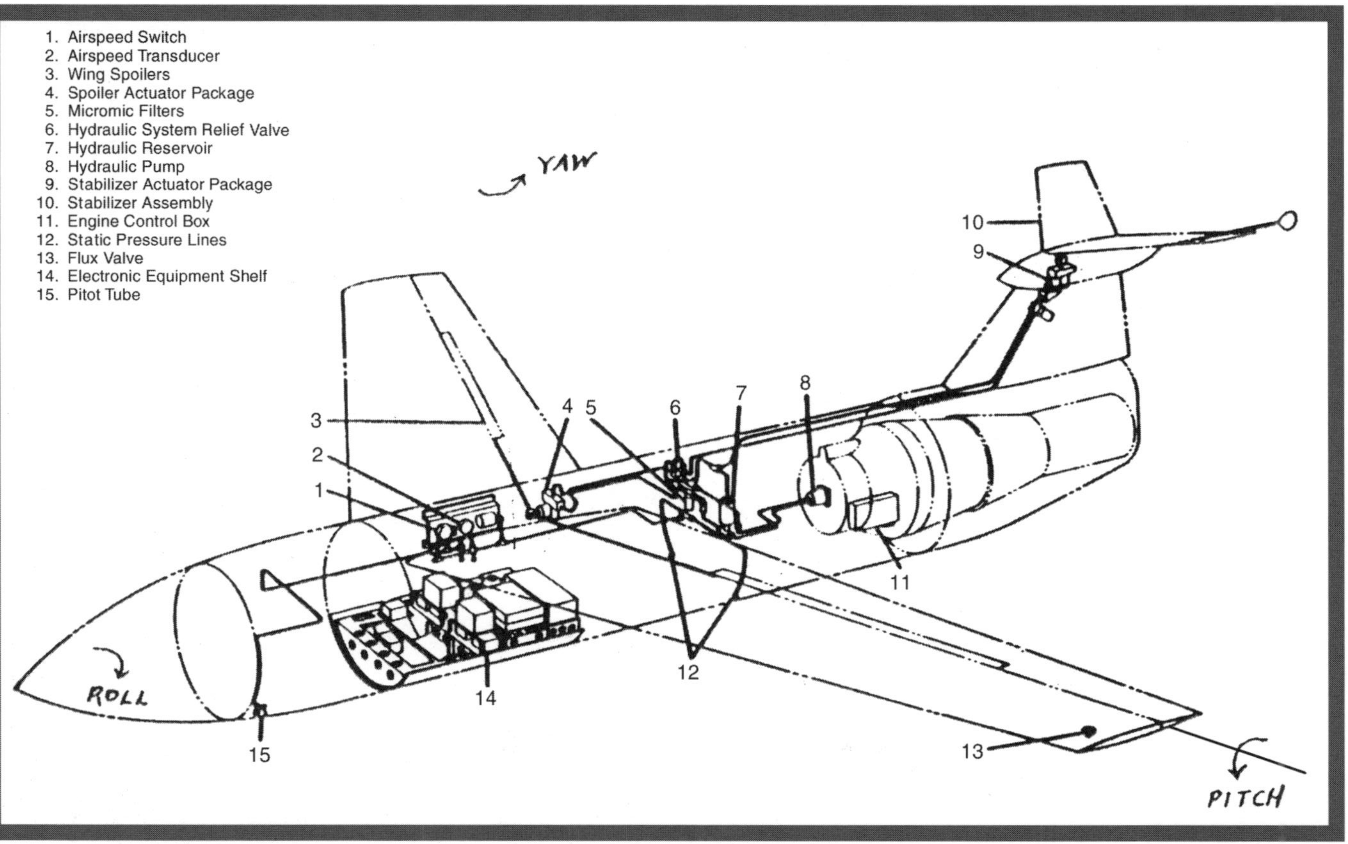

Figure 35. TM-61A flight control system.

Bomarc Missile, Boeing CIM-10A

The Bomarc (Figure 36), originally designed as the XF-99 and IM-99, was a surface-launched, pilotless, interceptor missile designed to destroy enemy aircraft. Propelled at launch by a rocket booster until it reached sufficient speed for its ramjets to operate, it was guided from the ground to target area at which time an internal target seeker took over the control. Testing of prototypes began in 1953 and the -A series was operational in 1960.

An improved -B series became operational in 1961 and had a range of 440 miles at a maximum altitude of 100,000 ft. It had a more powerful ramjet engine and its solid propellant booster allowed an almost instantaneous launch of a missile on alert. In 1969, Bomarc -Bs were operational at six USAF sites in the U.S. and at two RCAF sites in Canada. Bomarc -As were phased out in the mid-1960s, but beginning in 1967 some were modified and flown as supersonic high-altitude target drones. Complete phase-out of Bomarc's air defense mission was completed in October 1972.

General Data:
- Missile
  - Span: 18 ft. 2 in.
  - Length: 46 ft. 10 in.
  - Height: 10 ft. 4 in.
  - Weight: 15,600 lb
- Armament: Nuclear or conventional warheads
- Performance
  - Speed: 1975 mph
  - Range: 260 miles
  - Surface Ceiling: 65,000 ft.

F102 Armament, Falcon Missile

The missiles of the F102 are from the Hughes Falcon family of air-to-air missiles in the weapons bay and are AIM-4A radar-guided in the forward position and AIM-4D in the rear position. On the bottom center rail is an AIM-26A, a missile with a 1.5-kiloton nuclear warhead. Additional armament consists of twenty-four 2.75-in. folding-fin unguided Mighty Mouse rockets, two in each of the twelve tubes located inside the weapons bay doors. These rockets are partially visible at the forward ends of the doors. The

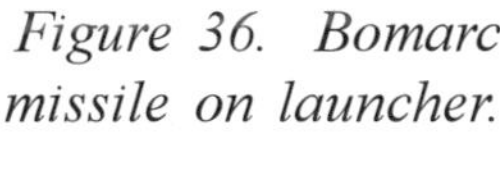
Figure 36. Bomarc missile on launcher.

weapons bays on both sides normally open together. To show both the open and closed configurations in this setup, only the right side is fully open to firing position.

Thc Falcon missile was stabilized in roll by a rate gyro. As the missile rotated, the various heat-seeking quadrants changed accordingly. Operationally the Falcon performed quite well. Technology gained on previous programs on heat seekers like the VB-6 Felix was useful in the Falcon designs.

AGM-62 Walleye

The Walleye I and II Air-to-Ground (AGM) Missiles were developed by Martin Marietta Corp., Orlando Division, in the early 1960s. They were unpowered

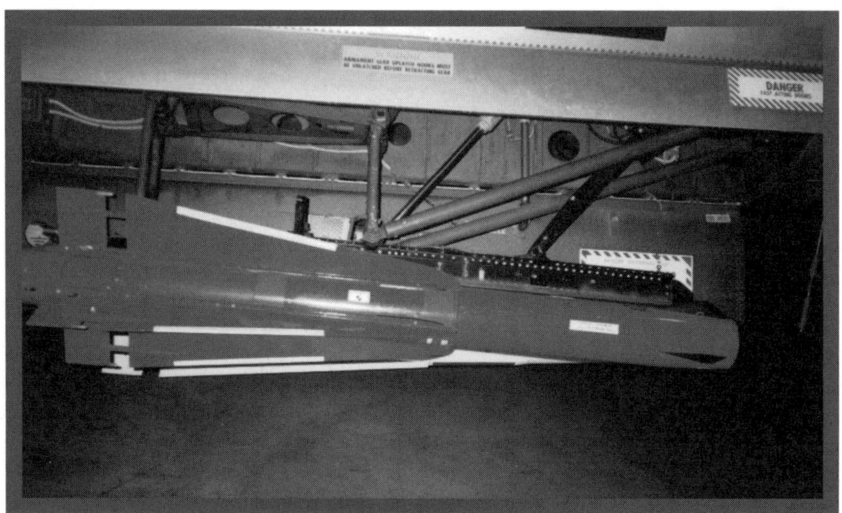

Figure 37. F102 Armament, Falcon missile.

glide bombs with a nose-mounted television camera. The bombs were used on the Air Force's F4 and the Navy's A4 and A7 aircraft during the Vietnam Conflict. The pilot would align the bomb with its target using a TV display in the cockpit. Once the bomb was locked on the target the pilot would launch the bomb and then he could either leave the area or launch another missile. During its gliding descent the Walleye could be controlled either by the pilot or by a pilot in a trailing aircraft. The pilot would monitor and update the bomb's aim point through a data link. The Walleye II was a larger version of the Walleye I and had a longer range.

General Characteristics

Walleye

- Length: 11 ft. 3 in.
- Diameter: 12.5 in.
- Span: 3 ft. 9.5 in.
- Weight: 1100 lb
- Range: 16 miles
- Warhead: 825 lb high explosives

Walleye II
- Length: 13 ft.
- Diameter: 18 in.
- Span: 4 ft. 3 in.
- Weight: 2400 lb
- Range: 35 miles
- Warhead: 2000-lb high explosives

Figure 38. AGM-62 Walleye missile.

Chapter 4

SLAM

SLAM is an acronym for Supersonic Low Altitude Missile. At the start of the program the SLAM utilized some far-out concepts based on the use of non-existent design technology. In particular it was to be supersonic at low altitude, just above treetop, powered by a nuclear ramjet engine and fly for months—and years—without refueling. Further, an aerospace vehicle of this type had never been built even at the experimental stage.

Design of Flight Control System

Obviously a number of assumptions were in order. It was in a sense like Columbus sailing to the new world; he had to assume the world was round or else he'd fall off the edge. On this program it was assumed a nuclear ramjet engine could be developed with the needed power for a guided missile that would meet mission requirements. It was further assumed that the flight control system, critical to missile flight, could be developed and ready for use at the same time the engine development had been completed.

Reviewing several space programs with their reentry knowledge it was felt the structure materials were for the most part available or could be without undertaking any extended research program.

The size and shape of SLAM was envisioned at the start along the lines of the B70. Of course a missile of this type would not have a pilot or need for a cockpit. Aerodynamic heating plus heat from the reactor would present a real temperature problem for all equipment, so a cooling system would be required for the guidance system and other items with temperature or nuclear or radiation limits. It was planned for such sensitive items to be installed in the nose or forward section of the missile along with any shielding that might be required. An artist's conception of the SLAM is shown in Figure 39. Numbers on the figure refer to the various components as listed.

Figure 39. Artist's concept of SLAM.

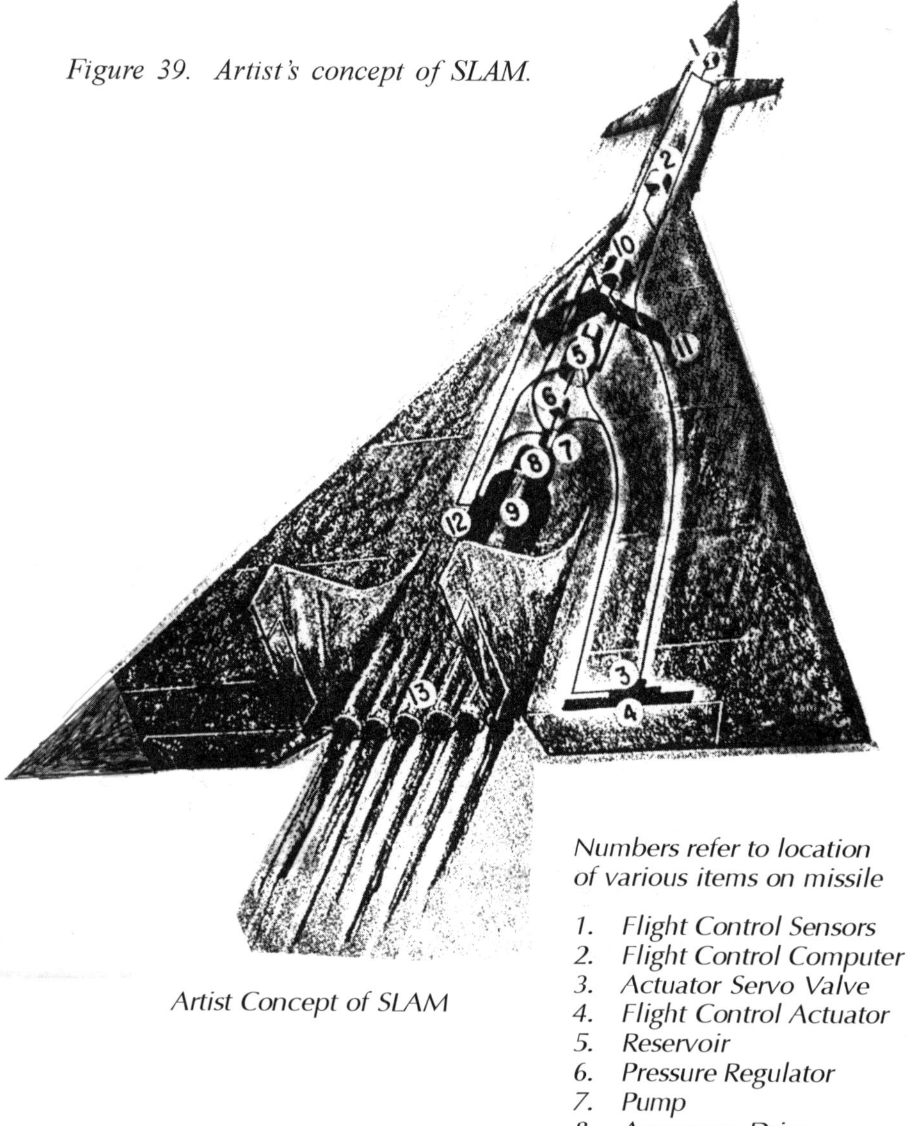

Artist Concept of SLAM

Numbers refer to location of various items on missile

1. *Flight Control Sensors*
2. *Flight Control Computer*
3. *Actuator Servo Valve*
4. *Flight Control Actuator*
5. *Reservoir*
6. *Pressure Regulator*
7. *Pump*
8. *Accessory Drive*
9. *Nuclear Engine*
10. *Alternator Power Package*
11. *Shielding*
12. *Nuclear Rod Contol*
13. *Nozzle Control*

Operationally it was planned that after launch, to be accomplished by a series of rocket boosters, the missile would go into a loitering phase, flying a figure eight at high altitude (approximately 80,000 ft.) over the ocean until it was called into use. It would then fly to enemy territory coming in at a low altitude so as not to be detected and at supersonic speeds in excess of Mach 3 to the target area or areas with a load of "A" bombs. This sounded like the best ever guided missile; all that was left was to make it a reality.

In short, it meant that to have an operational SLAM a number of firsts would have to be met: a nuclear ramjet engine would have to be developed; a flight control system that could withstand a nuclear radiation environment and high temperatures at or near 1000°F would have to be developed. These two alone meant a quantum jump in technology.

R&D programs were undertaken to provide flight control equipment that would meet these requirements. As will be described, some of which is in detail, technology was derived to build equipment for mission requirements. However, actual operational flight control design would have to wait until the vehicle design had been finalized and this, for the most part, would be changing only the size to meet installation or increased output requirements.

In addition to flight control design this same technology could be applied to other systems on the vehicle. In particular the actuation technology could be used in the engine gimbaling, called thrust vectoring, as it changes the engine direction of the engine output thrust, the position of the reactor fuel rod, and the position of the air duct for air entering the engine section. Figures 64, 65, and 66 in Appendix D show examples of these designs.

The "Flying Crowbar," as it was called, got its name from the Nuclear Ramjet Engine Program Director, Ted Merkle. This was apparently because of its low complexity and high durability. In a sense, one could say it was—or would be—built like a rock. Further, it didn't require a pilot and wouldn't be making rapid flight maneuvers, and the structural materials were available that could withstand the higher-than-normal temperatures. After all, it didn't require fuel tanks and numerous controls normally used on multi-engine aircraft. It was anticipated that a number of engines would be required on SLAM but controlling the output within usable limits did not appear to be an unsolvable problem.

The engine development for the SLAM was under the combined direction of the U.S. Air Force and the Atomic Energy Commission (AEC) and conducted at the Lawrence Livermore National Laboratory near Berkeley,

California. Static tests with full-scale reactor were conducted. To have a supply of ram air for these tests required 25 miles of 4B oil casing. The engine was mounted on a railroad car so it could be taken to a remote area 2 miles from the test site for disassembly and inspection, as it was radioactive after the test run. Final design ran for over 5 minutes producing 513 MW equivalent to 35,000 lb of thrust.

Although this design was successful, it appeared that a modified design might provide a Mach 4 vehicle. A number of problem areas also arose that required further consideration if additional engine development were undertaken, e.g., where and how to test new engines under flight conditions. Also, the Atlas and Titan missiles were being deployed at a price tag of $50 million each. After an extended review the government decided to terminate the SLAM project.

As with many projects that are never completed, some useful data and equipment are produced. Out of this program came the technology of how to design a ramjet nuclear engine and the development of a servo actuation system that would operate in a nuclear environment with fluid and ambient temperature of 1000°F.

The flight control system was next to the ramjet engine as far as developmental systems required for SLAM. In step with the engine development program, the Flight Dynamics Laboratory undertook the necessary R&D efforts to construct and test a servo actuation system that could be used on the SLAM and meet mission requirements.

During the next five years the General Electric Co., at their facility in Schenectady, performed the research, design, construction, and test on components and subsystem of the 1000°F actuation system. Diagrams of the subsystem are shown in Figures 55, 56, and 57 in Appendix D.

Initially it was not known what would present the most difficult technical problems, but now it might be quite useful to give a view of the system main component and an estimate of the technology needed to provide designs that could meet the SLAM mission as described at that time.

It was concluded at the start that the only design that could meet these stringent environmental conditions was a fluid power system which in turn defines the basic components and auxiliary types of equipment. A list of the basic items is as follows: a) the fluid; b) the pump or pressure device; c) the servo valve; d) the actuator; e) the feedback element; f) the construction materials; and g) facilities for evaluation of components and testing in environmental chamber.

Although the following write-up may seem lengthy and detailed, in many respects these are considered necessary in order that one can understand the problems that were involved and eventually the solutions that provided the technology and know-how to build an operational actuating system for the SLAM.

The fluid, commonly referred to as the hydraulic fluid, selected for the SLAM actuation system was NaK-77. This selection was based on results of previous R&D work in high-temperature fluids. However, what made this fluid so appealing was that it was currently being used in nuclear reactors as a heat transfer medium between the nuclear heat source and the cooling coils. A brief explanation of this fluid follows:

NaK-77 is a binary alloy of sodium (23%) and potassium (77%). It has a eutectic at this ratio of approximately 10°F; i.e., this mixture becomes a liquid at 10°F and remains so up to 1400-1500°F. It has a specific gravity of 0.87 which is about equivalent to the hydraulic fluid Mil-F-5606 then being used in military hydraulic systems. It had a bulk modulus that was higher at 1000°F (310,000 psi) than the oil was at room temperature. Bulk modulus is very important in flight control designs as it defines the actuator spring rate which in turn defines the flutter limits. Hence, the useful temperature range for NaK-77 was 10-1400°F. A separate study revealed that with the addition of cesium (Ce) the operating range could be –95 to +1300°F and this ternary alloy had similar properties as NaK-77. However, SLAM never had a low temperature requirement. Figure 60 in Appendix D gives some liquid metal properties.

In a nuclear reactor NaK was the only moving part. In an aircraft flight control system the fluid would be coming in contact with moving parts and different materials, so there was a problem of materials compatibility more in terms of temperature and friction or wear. The problem of containment was especially important because NaK would react violently with water or moisture. In fact all component testing with NaK was performed under a blanket of argon and in special chambers (see Figure 59 in Appendix D).

Lubricity of NaK was poor, therefore the use of positive displacement pumps, e.g., piston type, was ruled out and a dynamic pump had to be developed. A ten-stage centrifugal pump was designed such that each stage had an output of approximately 300 psi, for a total staged output of 3000 psi. This pump at rated speed of 30,000 rpm had an output flow of 12 gallons per minute at 3000 psi. A special high-speed drive (Figure 58 in Appendix D) was required to drive the pump at rated speed. Figure 61 in Appendix D shows an exploded view of the pump and Figure 62 the assembled pump as it was installed in the test fixture for evaluating the performance of the actuating subsystem at elevated temperatures. Materials used in construction of the NaK pumps included such metals as molybdenum and titanium carbide. The pump rotor assembly used borided molybdenum.

The following is a sample of pump test data: 33 hours at 500°F; 13 hours at 700°F; 5 hours at 900°F; and 1 hour at 1000°F. Two-vane pump designs were tested with NaK; however, both failed after 10-12 hours and before reaching temperatures of 500°F.

The actuator for the SLAM underwent a number of design changes before a successful component was available. In addition to seals material, compatibility was a problem. In the final design, parts and materials were as follows: barrel and piston - TZM molybdenum alloy; liner - tungsten carbide; piston rod - LW-1 flame plate; bearing - titanium carbide; seals - Inconel X; bolts - 442 stainless steel. The actuator had an overall length of approximately 27 in. at mid-position, stroke of 7.5 in., bore of 4.8 in., and a weight of about 45 lb. As with the B70 design, it was planned to have two actuators in parallel for each aerodynamic control surface. Assuming a 3000-psi supply pressure, each actuator would have an output force near 50,000 lb. Assuming four actuators were used to drive the elevator there would be a 200,000-lb force available for control. Two actuators were constructed in accordance with the philosophy of having two of each subsystem components available in case one failed; the second could be used to continue the test while the failed item underwent repair.

Materials compatibility on this program was a never-ending process over four years of development. Materials that showed good wear qualities at temperatures of 200°F would fail at the 800-1000°F range. Testing was accomplished by V-blocks on a Falex Tester. This tester was designed to test the specimens with NaK at normal and elevated temperatures in excess of 1000°F while varying the pressure on the test specimens.

The feedback transducer is a very important component of a control system as it reflects the system output to a command input. Further, it must be accurate, reliable with repeatable performance, and must operate over a wide temperature range.

Stated requirements were as follows:

Mechanical stroke: ±2.5 in.
Electrical stroke: ±2.375 in.
Operating temperature: 70-1200°F
Output gain: 0.03 V/in./V input
Linearity: 0.02%
Repeatability: 0.02%
Life at 1200°F: 15 hours minimum

Two feedback designs were purchased and tested; however, both failed at high temperature, i.e., above 800°F. Also, performance was not repeatable.

A study was made with the Research and Development Center of the General Electric Co. to design and build an LVDT (Linear Variable Differential Transducer) capable of 1000°F operation. A design and two units were provided to meet this requirement. An unusual feature of this unit was the use of a high parity alumina (Lucalox) coil bobbin shown in Figure 63 in Appendix D. The coils were wound with nickel-clad copper wire with boron-free glass insulation. Test results and operation of this transducer as a component of the servo actuating system at high temperature of 1000°F were quite satisfactory. Disassembly and inspection after tests showed this transducer design would meet all the operational requirements.

Servo Valve

Unquestionably the servo valve in any hydraulically powered control system is one of the most important components of the system. On this program it demanded considerable attention as no other system had such stringent requirements. The next important fact was that current servo valve designs just could not be applied. This was a natural barrier created by the temperatures involved. In particular, present servo valve designs used electromagnetics which found little value for this application. The operating temperature would be near 1000°F and all present designs would be faced with a Curie temperature. Curie temperature is defined as "a temperature at which the anomalies that characterize a ferroelectric substance disappear." This was found to be the situation as one servo valve that was supposed to be operable to 1000°F failed between 800-900°F. Also, performance was not repeatable. After three years of trying different designs, including poppet types, a fluidic design was found that met all the requirements. Inasmuch as fluidics is thought of as a new discipline, Appendix E explains how this design works and includes other related fluidic data. Figure 49 in Appendix C shows a schematic of the servo valve design.

Servo valve operation starts with an input signal to a summing element. If there is no other incoming signal, the original signal goes to an amplification section where it is amplified. This amplifier output changes the fluid stream direction coming from the jet pipe. The changed direction of fluid will enter one of the valve spool receiving ports; from there it goes to one end of the valve spool causing it to move. As the valve spool moves it uncovers a port which will then allow the fluid supply to flow to the actuator chamber to which it is joined.

The actuator piston is connected to a linkage that drives the airfoil surface. Attached to the actuator rod is a feedback device that sends a signal back to the summing section and, in turn, nulls the input. Simultaneously the valve spool has moved so as to have the receiving ports in a null flow position, thus hydraulic feedback is used in positioning the spool. When the input signal changes, say an amount equal and opposite to the original input, the operation takes place as before, only moving to the original position. The operation just described completes one cycle. Of course the travel of the actuator will depend on the magnitude of the input signal.

The feedback design used in conjunction with the fluidic servo valve used a design generally employed on the first stage of a two-stage nozzle-flapper valve—a bridge arrangement of two fixed orifices and two variable or changing orifices, the output of which drives the spool of the second stage. In the NaK servo-actuation system the variable orifices depend on the position of the actuator rod and, as this varies, it causes pressure fluctuations which are fed back to the input section of the servo valve and is summed with the input signal.

Test setup for the actuation system employed a number of specially designed components; e.g., just to heat the NaK to 1000°F took a minor loop which had to have a pressure source and this had to be near the main test arrangement. In short, there were a whole series of problems, some minor, but they all took time and resources. However, these problems could be solved with time and effort as compared to the designs for the system which was the real challenge.

The servo actuation subsystem represented in block diagram in Figure 40 was completely assembled and statically tested under various load conditions at fluid and ambient temperatures of 400°F, 750°F, and 1000°F in an inert environment. These static tests were performed to evaluate the steady-state positional accuracy, resolution, and repeatability of the system. Additional tests were conducted to ensure that all the system components, instrumentation, and auxiliary equipment functioned normally before making the dynamic evaluation of the subsystem.

The pump heater housing was replaced with an improved higher-temperature heating element (see Figure 41). Torsion bar assembly for load simulation is shown in Figure 42. The laboratory setup is shown in Figure 59 in Appendix D. In this photograph the Falex tester is in the chamber shown at the left. Here it can be placed in an atmosphere of argon and run at temperatures as high as 1200°F. The chamber shown in the background has the servo-actuation components that will be tested and evaluated at temperatures of 1000°F when operating with NaK. Argon is also used in this chamber as a blanket. At the far right in this figure is the NaK pump and the high-speed drive. During high-temperature tests a protective cover was placed over the pump as a

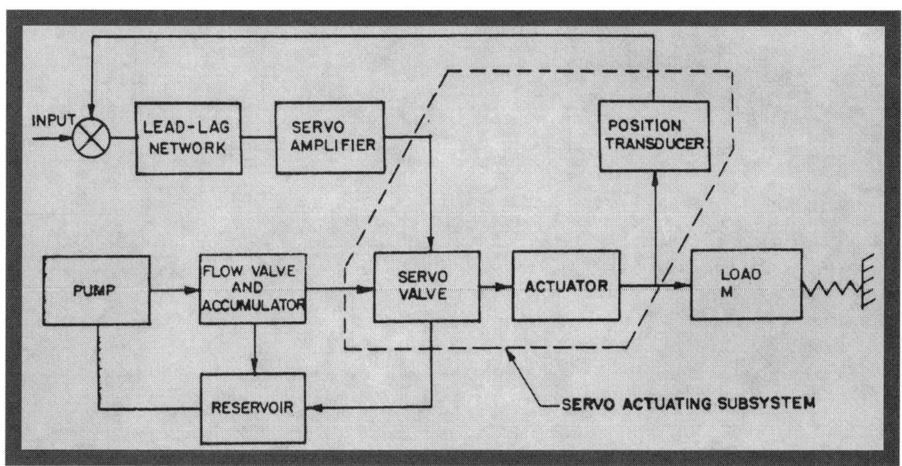

Figure 40. Servo-actuating subsystem.

Figure 41. Fluid reservoir with additional heating elements.

Figure 42. Torsion bar assembly for load simulator.

precautionary measure for the safety of operating personnel. A close-up of the pump and drive unit is shown in Figure 58 in Appendix D.

Dynamic tests were performed on the servo-actuating subsystem with NaK at elevated temperatures under varying load and input frequency conditions. As it required over an hour to bring the fluid temperature to 1000°F (actually 1002°F on instrumentation meter), a series of tests were conducted at 400°F and 750°F. Rated mass load on actuator was 45 lb-sec^2/ft. and maximum load torque was 8 x 10^4 in.-lb. At a load amplitude of 15% and temperature of 1000°F the subsystem was operated from 0.5 Hz to a frequency of 20 Hz (Hertz = one cycle per second).

A sample of the subsystem performance for this test is shown in Figure 44. It shows the resonant frequency of the closed-loop response to be about 7.5-8 radians/sec. In terms of time, the subsystems operated for 1 hour at 400°F, 1.5 hours at 750°F, and 1.5 hours at 1000°F; thus the total test time was a minimum of 4 hours of continuous operation.

During these tests the equipment was enclosed in a controlled temperature oven and the whole assembly covered by an inert atmosphere of argon.

In summary, dynamic testing and evaluation of test results demonstrated that a liquid metal (NaK-77) could be constructed that would meet the stringent environmental conditions required of the flight control system for a SLAM.

Test results based on instrumentation data:

Servo actuation components consisting of pump, actuator, servo valve with fluidic amplifier and feedback transducer were exposed to the 1000°F environment.

Pump: 30,000 rpm, 8 gpm, 3000 psi

System Response: 3db at 2.5 Hz, input amplitude 5% of max

Actuator travel: ± full stroke 100,000 in.-lb output

Figure 43 shows a conventional valve-actuator so that one can see the magnitude of changes required to produce a design capable of 1000°F and nuclear environment. The actuator of Figure 43 is a schematic of a design that was exposed to a radiation environment for 200 hours. Tests were conducted at periodic intervals to determine change in performance. The hydraulic fluid used on this test was the standard Air Force Hydraulic Fluid Mil-0-8200. All tests didn't determine what materials one must use for application in a radiation environment but it clearly demonstrated what not to use.

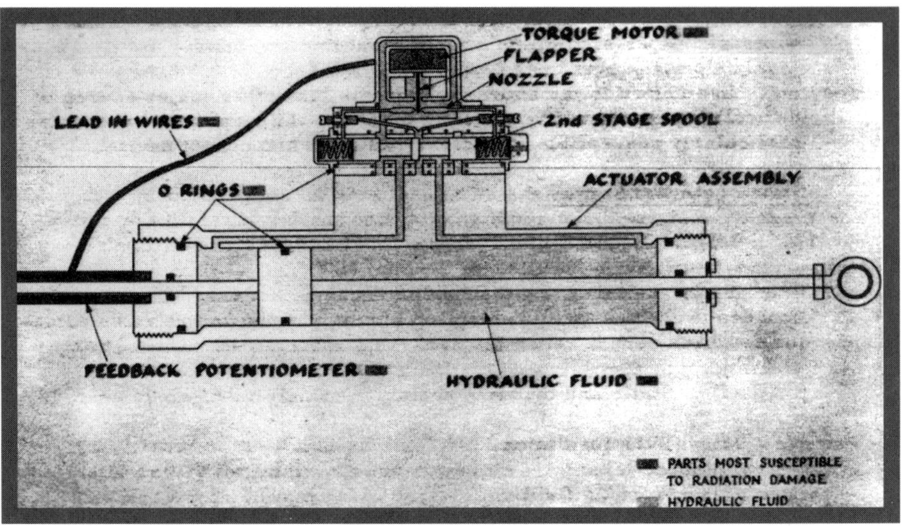

Figure 43. Parts of valve actuator assembly most susceptible to radiation damage.

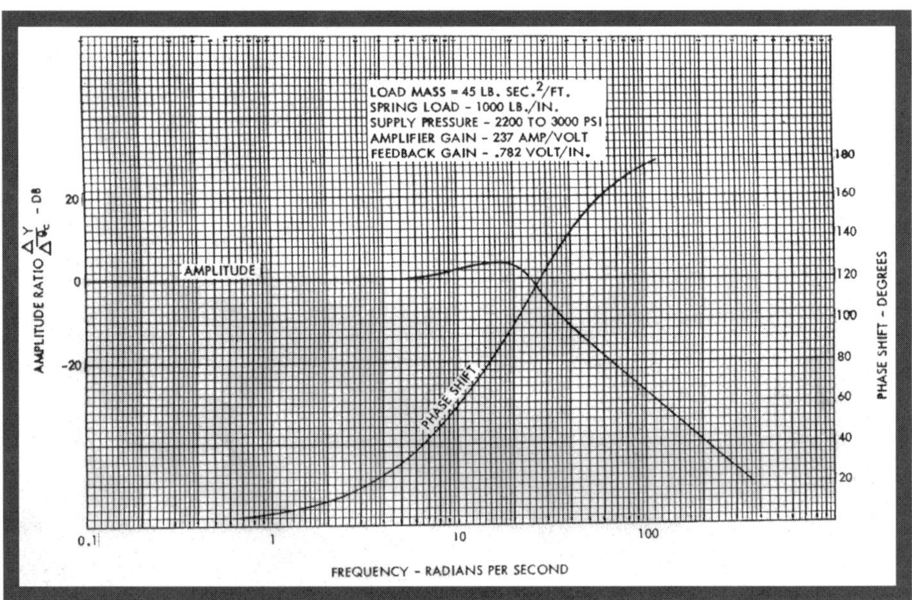

Figure 44. Graph of closed-loop frequency response liquid metal servo valve actuator.

SUMMARY

Reviewing this program at this date shows that a quantum leap was made in terms of what the R&D accomplished. Namely, establishing the technology to design an actuation system capable of operating at 10-1000°F in a nuclear radiation environment. In terms of cost, this entire program over a five-year period had four contracts (including three subcontracts) and the cost was in the range of $700,000-$800,000. Apparently there isn't a present need for such a system but surely as we get into space travel and unknown conditions a future need is very likely to occur. Therefore, unless some group follows up on this work it will probably end up like the Egyptian mummies and be buried for centuries without any known useful purpose.

Chapter 5

Controlled Aircraft

Drones and targets fit into the category of controlled aircraft. The general concept is that drones are derived from operational aircraft whereas targets are designed and constructed from scratch. Using such a classification the Controlled Aircraft Unit, Flight Control Laboratory, had a number of projects to convert existing aircraft into drones, including the B-17, B-29, and P-80.

Drone Aircraft

Drone aircraft have never shared the limelight as have some military aircraft that demonstrated their performance capabilities in combat. However, as a means of gathering crucial data or being destroyed as a target, the drone was indispensable. In many respects, drone aircraft can be thought of as aerospace robots.

B-17

The B-17 (Figure 45) was the first large aircraft to be converted to a drone. B-17s were converted to drones as early as 1944 and a series of tests were conducted at the Wendover Air Base. The touch-and-go tests were next to spectacular, as the aircraft appeared to be under perfect control by a pilot.

Changing an operational aircraft to a drone meant removing the armament of guns and similar types of equipment that would not be required. It was called a stripped-down version of a B-17. Flight control of the aircraft was dual, i.e., a pilot could fly the aircraft or it could be remotely controlled from a station outside and some distance away. The C-1 autopilot was modified, as were a number of items, in order to respond to the radio signals from the controller. During the A-Bomb tests in 1947 at Enewetok in the Pacific Ocean, a B-17 drone was instrumented to gather data on the radiation levels. The man in our group responsible for the B-17 drones was sent to monitor and maintain the aircraft during these tests, which lasted about three weeks. We never knew if he got too much radiation at that time, but some years later he took early retirement because of physical disabilities.

Figure 45. B-17 drone.

B-29

There were two significant efforts that were completed and demonstrated on the B-29 drone (Figure 46) by our group as a test bed. The first of these described was an "in-house" effort; i.e., equipment was designed and fabricated in our shops and installed on the aircraft by our own mechanics.

The nose wheel steering in the normal or standard installation required the pilot to move a hand lever. Nose wheel angle was changed by a hydraulic actuator with the pilot controlling the valve that ported hydraulic fluid. A modification was made that allowed the nose wheel steering to be interconnected with the rudder system. This system also had a pilot override, i.e., an added element that allowed the pilot to take over during test of the system before becoming a drone system. A drone requirement on this system concerned the takeoff procedure. The drone at takeoff would be steered down the runway by the nose wheel steering. Once liftoff speed was reached the

Figure 46. B-29 drone.

rudder would take over so the transition between the mechanical and aerodynamic control was quite smooth. A second part of this design allowed the pilot to taxi the aircraft with the rudder pedals, which most pilots liked because it permitted them to have a free hand to move the throttles at the same time. This aircraft used a Sperry A-12, E-4 autopilot.

A second development on this project concerned the throttles. A design by the Bendix Corporation called the "Uni-Lever Power Controller" moved all four engine throttles with one lever. It had a number of electronic components and, once calibrated and working properly, its operation was fairly reliable. In fact, the system was so designed that when the four engines were generating a near equal amount of power per engine, the Uni-Lever locked all four with an arrangement that the single lever moved all four together. This was also a drone requirement as it would have been a real problem if the engines were controlled separately. Hence, technology gained on the Uni-Lever design could be used on other multi-engine aircraft.

P-80

The P-80 (Figure 47) was converted to a drone and flight tested. The actual drone testing was conducted at a more or less remote location as a safety measure, as it was the first P-80 to become a drone. Flight test results given to our group said it performed quite well. This aircraft, like the B-29, used a modified A-12, E-4 autopilot. Actuators for moving the controls were the same type and design used in the VB-3 bomb.

Figure 47. P-80 drone.

Chapter 6

Summary and Conclusions

From the early programs on controlled bombs (1944-45) came a number of guided missile programs after WWII which, in turn, helped make the United States military the world leader in this field.

The VB-3 proved by operational use to be a very effective weapon. If the developmental phase had been completed and the bombs kept on a "ready" basis it undoubtedly would have shown more fruitful results.

Reviewing the R&D programs with the General Electric Co. at this date shows:

- That the technology to design, construct, and test a flight control actuation system capable of operating from 10-1000°F in a nuclear radiation environment had been developed;
- In terms of costs, the entire program over a five-year period with four prime contracts and three sub-contracts was accomplished for a total of $700,000-$800,000.

The use of drone aircraft proved invaluable in obtaining radiation levels in an area radiated by a nuclear blast immediately after its detonation.

As with the story of the Sinking of the Rhona (Appendix A) which was held from public view for over 50 years, I feel that all those who supported our WWII efforts without knowing what was taking place on new weapons are now entitled to be informed how and what was produced for their dollars. They also should see how much was accomplished in such a short time with limited personnel and resources.

If ever there is one point to get across to the uninformed, it is that unless research is continually being conducted to advance weaponry, we will not be prepared in time, as was so clearly demonstrated in WWII. General Arnold recognized this and did his best to enable us to catch up.

References

1. AFFDL-TR-68-82, *Study of Liquid Metal NaK-77 for Application in Flight Control Systems.*

2. WADC Technical Report 57-294 Part III, *Research on Liquid Metals as Power Transmission Fluids.*

3. AFFDL-TR-64-183, Vols. I and II, *Research and Experimental Investigation of Liquid Metal NaK-77 for Application in Flight Control Systems.*

4. ASD-TDR-62-597, Vols. III and IV, *Study of a Liquid Metal, NaK-77, for Application in Flight Control Systems.*

5. AFAPL-TR-65-73, *Multi-Component Alkali Metal Alloys.*

6. Gulf Research and Development Company, Progress Report April 1, 1945, *Experimental Investigations in Connection with High-Angle Dirigible Bombs.*

7. Handbook for Vertical, Azimuth and Range Controllable Bomb Tail, Type B-1, T.O. No.11-5G-4, 14 February, 1946.

8. VFW Magazine, October 2000.

9. Fortune, Don, *Sinking of the Rohna, America's Worst Troopship Loss in World War II,* published by the author.

10. Liquid Metal Handbook - Sodium - NaK Supplement TID-5277.

11. NASA TN-D-3928, *Operation of Hydrodynamic Journal Bearings in Sodium at Temperatures to 800°F and Speeds to 12,000 rpm.*

12. Hoyt, Edwin P., *The Kamikazes,* Arbor House Publishing Co., New York, NY.

13. Controlled Aircraft Group, Flight Control Laboratory, *Daily Activity Report 1948.*

Appendix A

"Pioneer to the Rescue"

Excerpts from a story printed in the October 2000 issue of the VFW monthly publication, titled "Pioneer to the Rescue" by Don Fortune, have been included here because they show the devastation caused by a controlled bomb (reprinted with permission from Don Fortune).

> **In November 1943, off the coast of North Africa in the Mediterranean Sea, the crew of the *USS Pioneer* saved the lives of hundreds of men from the bombed troopship Rohna.**
>
> At 4:40 p.m., the *Pioneer* sailed 37.06 North, 5.56 East, in the Gulf of Bougie, 15 miles off North Africa, guarding P Sector on the convoy's port beam.
>
> Suddenly, three *Luftwaffe* squadrons roared out of the sun without warning. Heinkel 111s and 117s, Dornier 217s, Junkers 88s, and Focke-Wulfs swung wide of the convoy, concentrating fire on the escorts.
>
> Several aircraft carried highly secret, rocket-propelled, remote-controlled Henschel Hs-293 glide bombs. Winged like planes, the air-to-surface missiles packed 1,100 pounds of explosives.
>
> Bombs fell as *Pioneer* gun crews sprinted to battle stations.
>
> Scrambling from North Africa Bases RAF Squadron 153, USAAF 350 Fighter Group, and the Free French 1st Fighter Group engaged the Germans.
>
> *Pioneer* Capt. LeRoy "Cowboy" Rogers directed action from the flying bridge. Gunners sent up a blistering barrage from port and starboard 20 and 40 mm guns and from the 3-in./.50-caliber gun on the bow.

> Boatswain's Mate 2nd Class Wayne Dana, gun captain on a 40mm gun, "was ready to fire on what looked like a small plane, but somebody said, 'don't fire, it's a bomb.' It's a good thing I didn't fire, because it would have burst in air. It jarred the ship, and three men in our crew were hit, but recovered..." . . . .
>
> . . . . The *War Diary* reported the ship fired 59 rounds of 3-in. shells, 222 rounds of 40mm, 950 rounds of 20mm, and 150 rounds of .30-caliber machine gun fire.
>
> **ROHNA TRAGEDY**
>
> Close by, the British transport *Rohna,* with almost 2,000 American soldiers aboard, ran out of luck.
>
> A Heinkel 177 bombardier targeted *Rohna* with a secret missile—Hs-293. The bomb fell behind the mother plane, overtaking it as its rocket ignited. The nose glowed red; fire blasted from the tail. Moving the joy stick, the bombardier swerved the bomb into *Rohna's* port side. It exploded in the engine room, gouging a truck-sized hole on both sides of the ship.
>
> *Rohna* lost all power, caught fire, and began listing heavily to starboard. Hundreds died instantly. The 853rd Engineer Aviation Bn. lost 67% of its force. By nightfall, 1,015 GIs had died, the greatest loss at sea of U.S. Army personnel in WWII. Three American Red Cross men, five *Rohna* officers and 115 native crewmen died also . . . .
>
> . . . . The overladen *Pioneer* finally turned her bow toward North Africa at 1:45 a.m. on Nov. 27, 1943, after the last survivor sighted was brought aboard. The ship sailed flank speed for six hours until reaching Philippeville (Skikda) in Algeria. There, 601 survivors and five dead were disembarked . . . .

Description of the Hs-293 bomb is given in Chapter 2 of this book. The Germans had a 5- to 10-year start over the U.S. when it came to guided bombs (or missiles). This was clearly demonstrated in the sinking of the *Rohna* in 1943 when they used a powerful glide bomb (Hs-293) and employed a "Line of Sight" technique for guiding the bomb. We should note

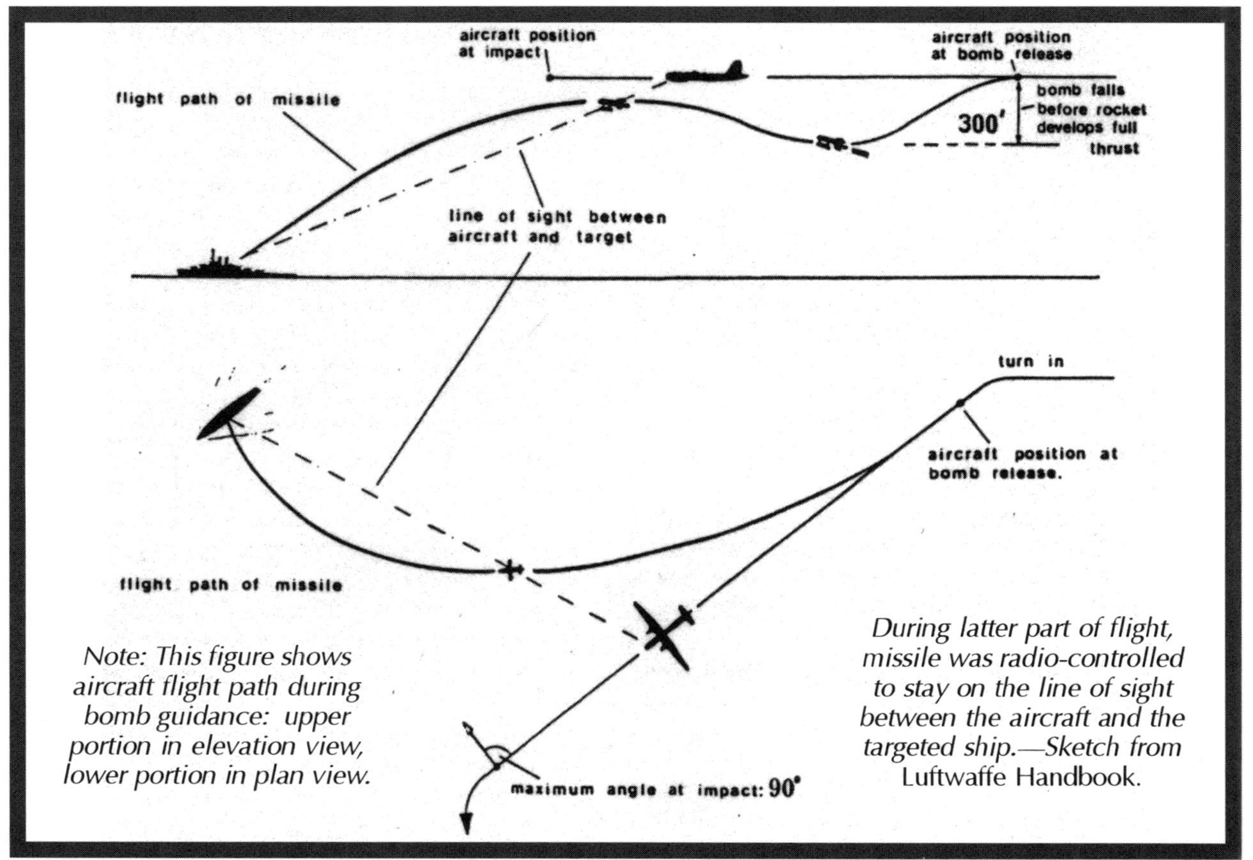

Figure 48. Hs-293 line of sight diagram.

also they had a small engine on this missile which would be necessary to get the missile ahead of the aircraft if any form of "line of sight" was to be employed. They probably hadn't perfected any "line of sight" for missile control but apparently they were on the right track. Obviously to use such a method they must have started working on it at least five years earlier.

What all this shows is that the research is required years in advance if one desires to have useful weapons for operational use.

Appendix B

Technical Notes

The transition from controlled vertical bombs to guided missiles involves the use of technology that came into being during the period of 1943-93. The vertical bomb was, in many respects, a follow-up of the ballistic shell delivered by gunfire or, in the case of large shells, by artillery. Further, the trajectory relied mostly on the shape of the shell, and the angle and velocity from which it was fired. Of course rotation, or spinning the shell in flight, allowed it to be more stable with a more predictable trajectory. Little or no thought was ever given to the possibility of in-flight control. Hence, with the controlled bomb, not only was control involved but thought was given to modifying or changing the shape of the bomb and, in most cases, adding surfaces which now meant that aerodynamics and all of its technical background would enter into the picture. As wings were added the bomb (the warhead) now became a part; also, except for glide bombs, an engine or sometimes a rocket was included.

As one might expect, it was paramount to the mission's success that these new weapons flew a given path and maintained stability at all times. While aerodynamics came into play, modern control theory was being applied to the control members and eventually we had the development of the flight control system. Modern control theory provided the analytical means to design flight control systems to keep the guided missile stable under all flight conditions. This was quite a step forward as many a missile was lost in flight due to an unstable flight condition.

Some people called guided missiles "Glorified Robots," and to a certain extent this was true, especially for the designs in which the guidance system was pre-set and took the missile from launch to target. There are types such as ICBMs (Intercontinental Ballistic Missiles) where the missile is rocket-launched and put on a trajectory that will take it to the target. In short, once it is up to speed and on a fixed trajectory, it assumes the ballistic trajectory.

After WWII, a number of the German engineers were assigned to our group at Wright Field. They explained several design techniques that were quite interesting. In particular was the control method to provide moments on the V-2 to keep it stable during the initial launch phase. The accepted procedure in today's designs use engine gimbaling, also referred to as "Thrust Vectoring." However, such techniques had not been developed in 1945. What they did was use a series of oak (wooden) airfoils in the engine thrust jet. These were also connected to the larger external aerodynamic fin and, as these wooden airfoils burned up, they no longer produced any moments and the external fins took over control.

At times the operation of equipment systems is written in the vein that everyone completely understands "how they work" or "why they work like they do," so it might prove worthwhile to cover several items as a matter of interest. For example, a control system has a sensor, the computational electronic circuitry where the inputs are added, subtracted, multiplied, etc., to provide an output to the actuation system and it, in turn, moves the aerodynamic control surfaces. Attached to one of the output members is a feedback transducer that sends a signal to cancel the sensor signal. The process is repeated in the opposite sequence as the missile returns to its original position.

Sensing Elements

Just as a magnetic compass gives us direction, or an oriented reference in a plane parallel to the earth's surface at a given point, a gyro can provide a directional reference in space. The gyro, which is essentially a wheel-like mass rotating at high speed, accomplishes this by putting it into a set of gimbals that are free to rotate. The gyro spin axis will remain in a fixed position as the other members are rotated. Attached to the gimbals are electrical transducers that produce a voltage output in proportion to the amount of gimbal angular motion. It is the gyro sensing design that is used on most missiles in order to maintain the desired orientation for any given flight condition. The output of other sensors, e.g., airspeed and altitude, are also used to control the missile.

Computation Section

Signals from the various sensing elements are combined in a section, usually electronic, where they are added, subtracted, or altered to provide an output to the next component. Further, amplification is used in order to increase the output magnitude that is required for the next component. In a hydraulic actuation system this component is an electrohydraulic servo valve. In the flight control field many individuals refer to the computation section as the "electronics," where lead-lag networks perform the functions that produce a desirable output for the next component in the system.

Actuation Section

The actuation section, be it pneumatic, mechanical, electrical, or hydraulic, provides the power to drive the control output which is usually connected to an aerodynamic control surface. Also attached to an output member is a feedback transducer that sends a signal back to the computation section where it is combined with other incoming signals to produce the required output.

The above descriptions are very elementary to the technical individuals in the aerospace business but were included here to give others just a general idea of what is occurring in one part of the guided missile operations.

Appendix C

Fluidics

Fluidics is a discipline that uses fluid flow and specially designed elements to perform sensing, logic, amplification, and control functions. It is only about 35-40 years old. Webster's 10th Edition 1995 defines *fluidic* but not *fluidics*. *Fluidic* is treated as an adjective and defined: of, relating to, or being a device (as an amplifier or control) that depends for operation on the pressures and flows of a fluid in precisely shaped channels. *Fluidics* is a noun but has no general definition.

The SAE uses the following definitions:

> *Fluidics*: The general field of fluid devices and systems performing sensing, logic, amplification, and control functions employing primarily no-moving-part (flueric) devices.
>
> *Flueric*: An adjective which can be applied to fluidic devices and systems performing sensing, logic, amplification, and control functions if no moving mechanical elements whatsoever are used.

Thus we conclude flueric devices are a group of fluidic devices without any moving parts.

Of course we must remember that in all these devices, fluid is flowing, and as the input pressure is changed the flow and pressure in other sections of a device is altered. How this is accomplished depends on the design shape; hence, the location of the output port defines the flow and pressure changes due to input change.

Figure 49 is a schematic of a flueric element. Fluid enters at the chamber marked SUPPLY, it then flows through the element such that the flow is equally divided and exits at the OUTPUT FLOW. The dotted line in the middle of this element is the magnetic field set up by the permanent magnet. Now if an electric current is sent through the electric coiled lines the flow is

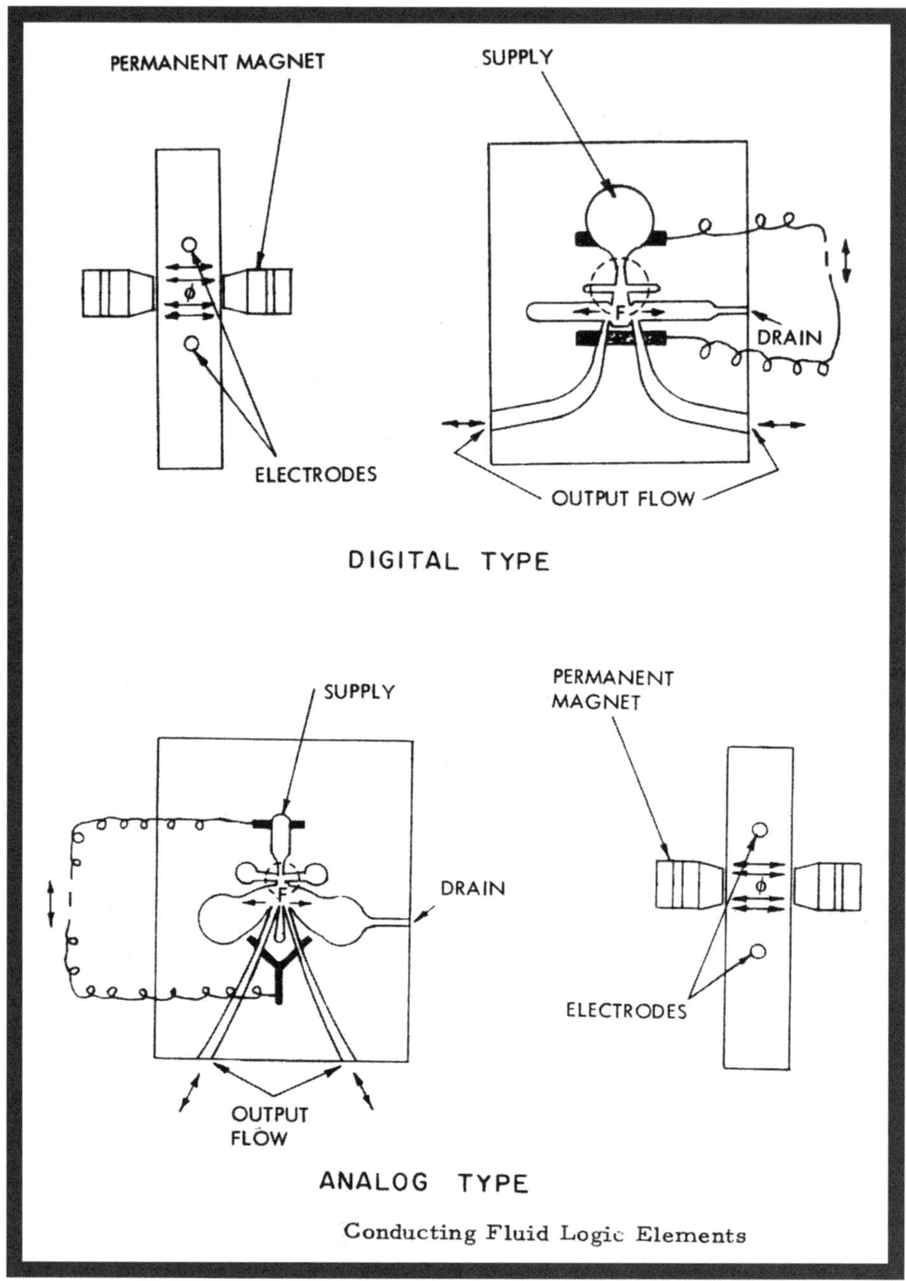

Figure 49. Schematic of a flueric element.

no longer equal to output channels. This flow is a function of the magnitude of the current and the difference in flow is used to drive another element in another device. Hence, with proper design the various flows and pressure differences can be used as in any mechanical design. The direction of the magnetic field and current direction dictates the direction of the force or fluid pressure. This is sometimes referred to as the left-hand or motor rule.

Figure 50 is a photograph of the flueric element just described. It has been blown up for greater visibility; however, size-wise it would be about the size of a postage stamp. The shiny element in this photo is the permanent magnet.

Figure 50. Photograph of flueric element.

Rapid progress has been made within the past several years in the development of fluidic devices for logic, sensing, and computational purposes. Flueric devices have the inherent advantage of no moving mechanical parts, and can be made relatively independent of temperature variations. In many cases both the input and output intelligence are fluid so that conversion of energy from one form to another is eliminated. In other cases, however, electrical signals must be accepted, and here recourse must be had to the conventional electrical-mechanical energy conversion. When, however, an electric conducting fluid such as a liquid metal is used, it becomes possible to employ a unique control concept that eliminates the mechanical energy conversion. The operating principle utilizes Ampere's law; i.e., the force on an electrical conductor is proportional to the current, magnetic flux density and length of the conductor in the magnetic field. When the conductor is a stream of conducting fluid it will be deflected in a direction and manner analogous to the rotor in a conventional electric motor. A fluid logic device can thus be designed to provide proportional fluid output with no moving parts and with very simple electrical and magnetic circuitry.

Since the pressure gain of such an element is very low, additional hydraulic amplification is necessary to control the second-stage spool of the servo valve. This amplification is accomplished by using a five-stage flueric amplifier. Using these concepts with a liquid metal as the hydraulic fluid makes it possible to combine functions of the servo valve torque producer and pilot stage in one unit, and at the same time provide hydraulic feedback both between the two valve stages and between the controlled load and the servo valve.

In operation the electrically controlled flueric element deflects a jet of NaK fluid to provide a proportional input to the multi-stage flueric amplifier. The output of the amplifier deflects a fluid jet that is directed into receiver ports in the second-stage spool. Hydraulic feedback provides a one-to-one correspondence between jet displacement and spool position.

Motion of the valve spool uncovers a supply port and allows fluid to flow to the actuator, causing it to move at a given velocity. The hydraulic feedback device attached to the piston rod has two fixed and two variable orifices. The latter are adjusted so that they are equally distant from the linear cam when the piston is at mid-position. Under this condition fluid flowing through the fixed orifices and out the nozzles produces equal pressures in the chambers between fixed orifices and the nozzles. These pressures are fed back to the input element; however, when they are equal there will be no net effect on the direction of the power jet and feedback signal is zero. Figures 51 and 52 are schematics of a fluidic servo valve and hydraulic feedback device. Figure 53 is a photo of a feedback device installed in an actuation system (Figure 54 is an exploded view).

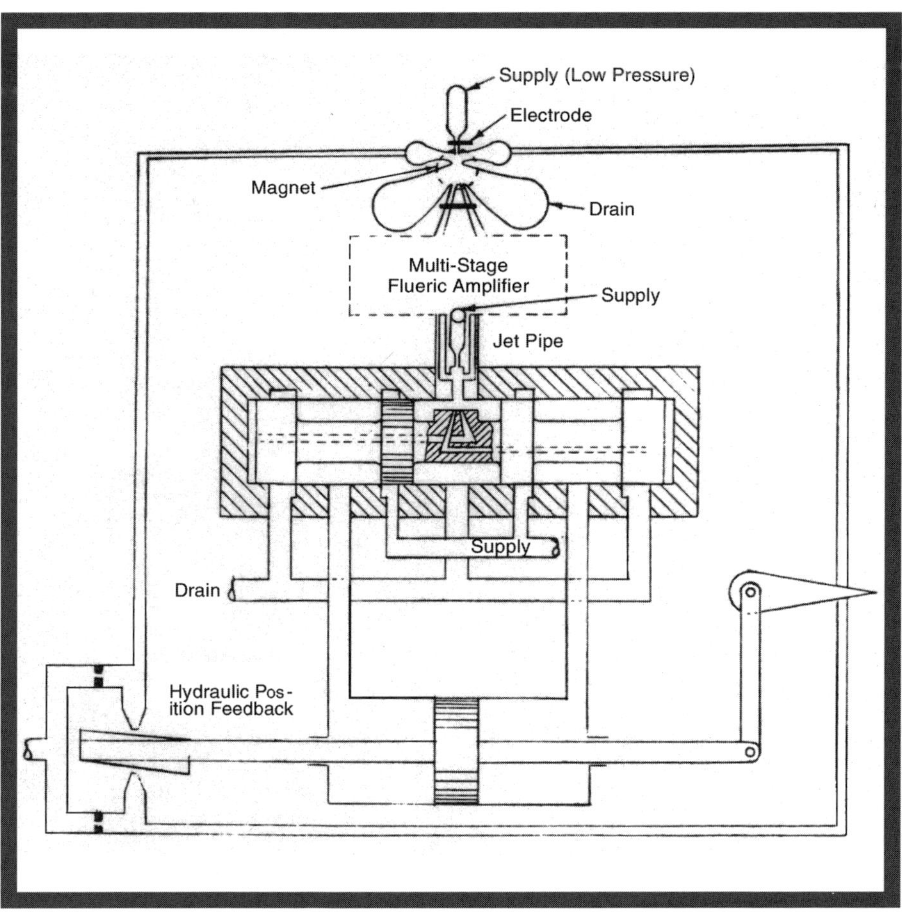

Figure 51. Schematic of a feedback device.

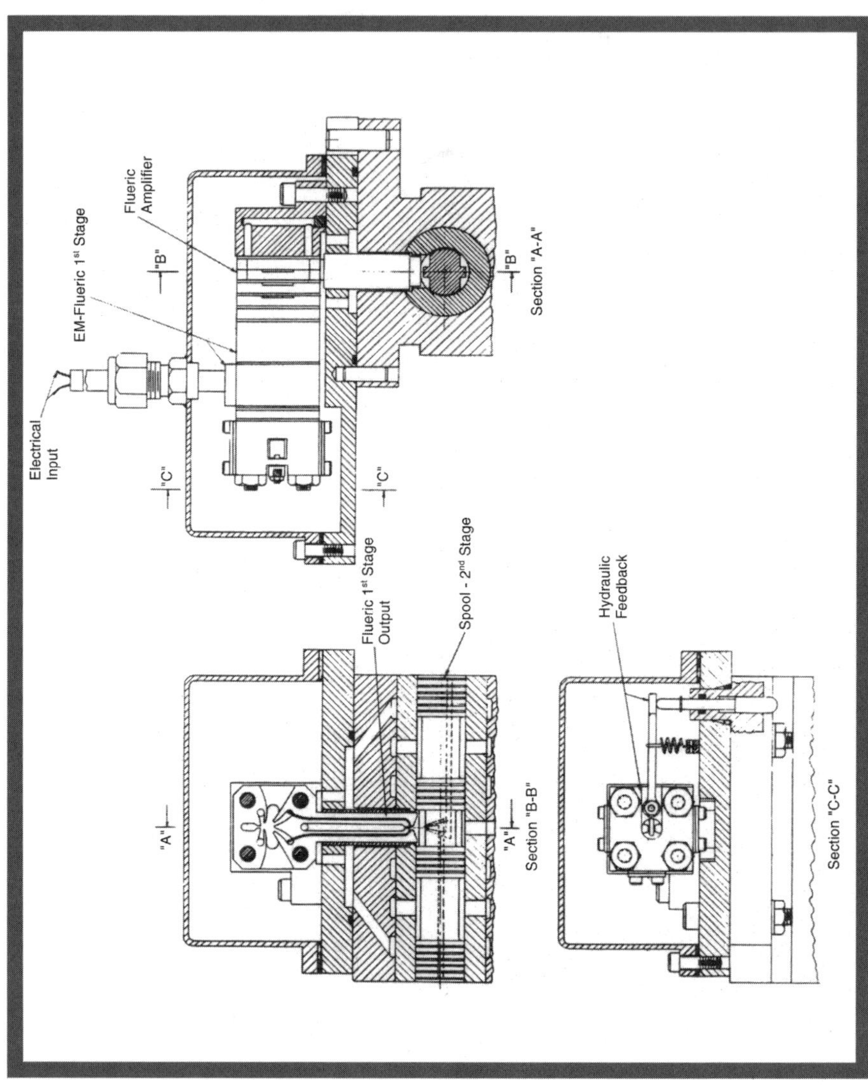

Figure 52. Fluidic servo valve and hydraulic feedback schematic.

Figure 53. Photo of feedback device.

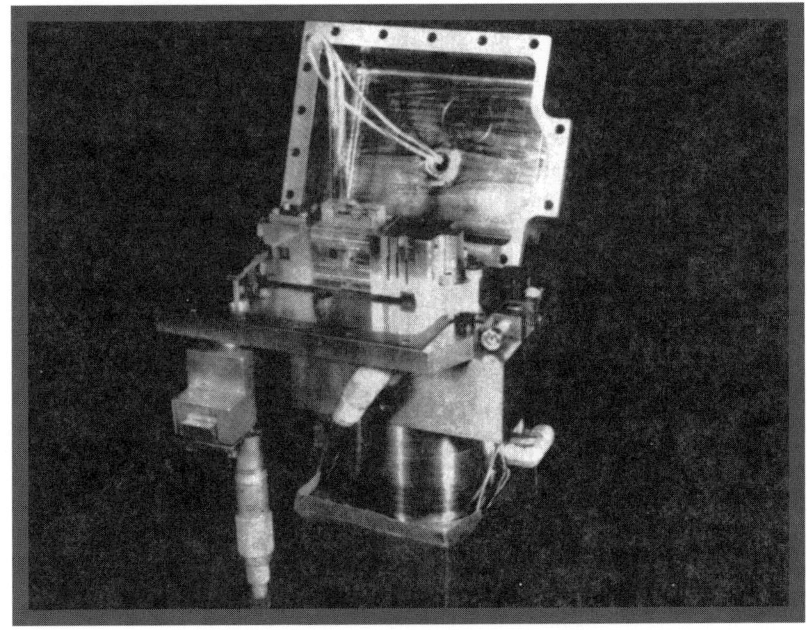

Figure 54. Exploded view.

Appendix D

NaK Facility and Components

On the following pages are photos and figures related to the NaK Flight Control Actuation system. All work and programs undertaken to determine the use of liquid metals in flight control systems were performed by the General Electric Co. at its Liquid Metal Facility in Schenectady, New York, for the Air Force Flight Dynamics Laboratory under direction of V.R. Schmitt, Project Engineer.

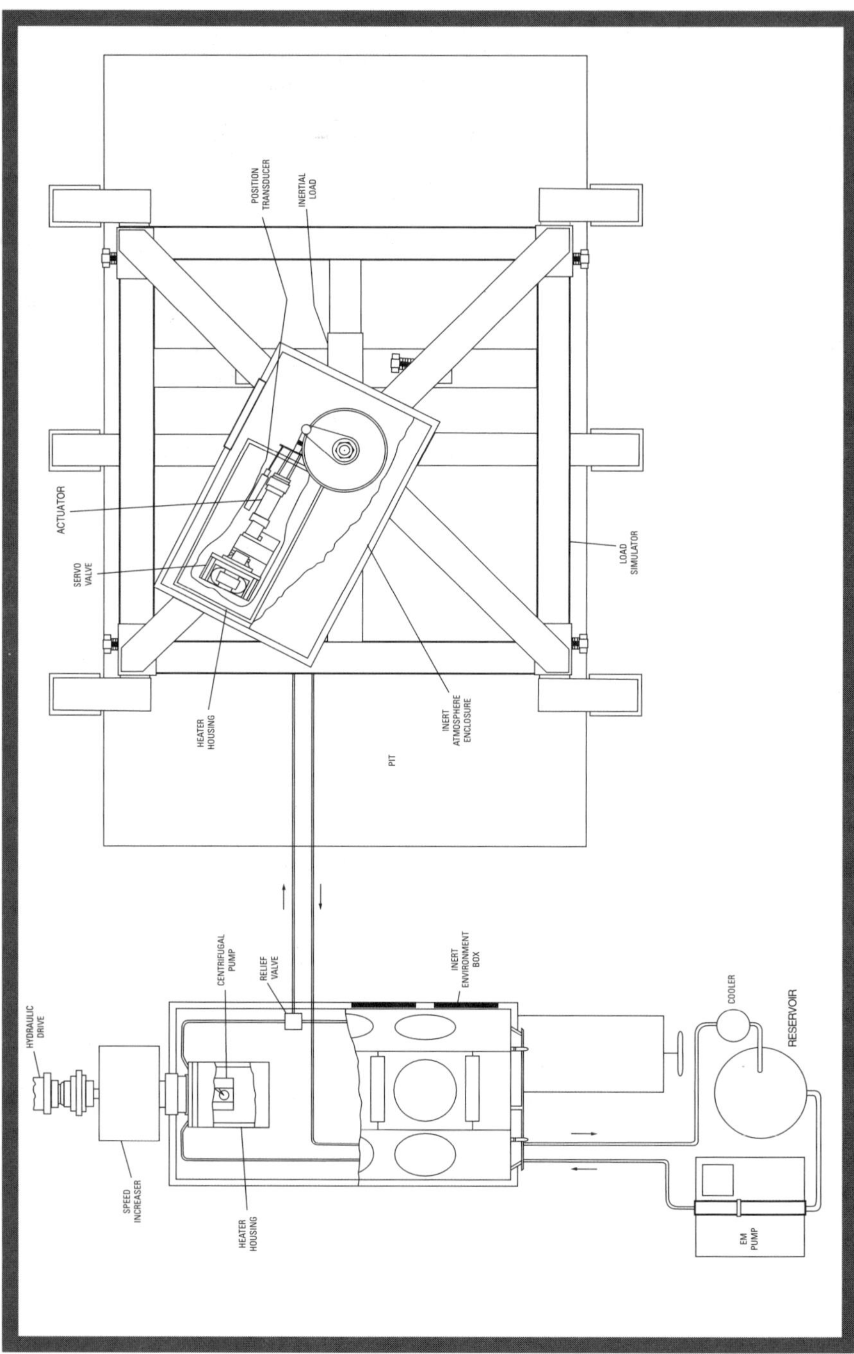

Figure 55. Schematic of servo actuating subsystem.

Figure 56. Actuating subsystem on load test stand.

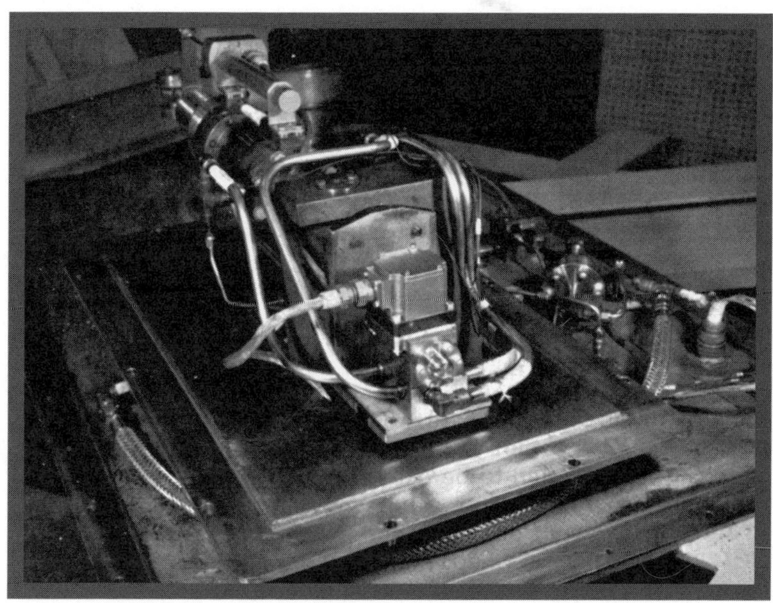

Figure 57. Close-up of actuator on test stand.

Figure 58. High-speed drive and pump connected to inert atmosphere glove box.

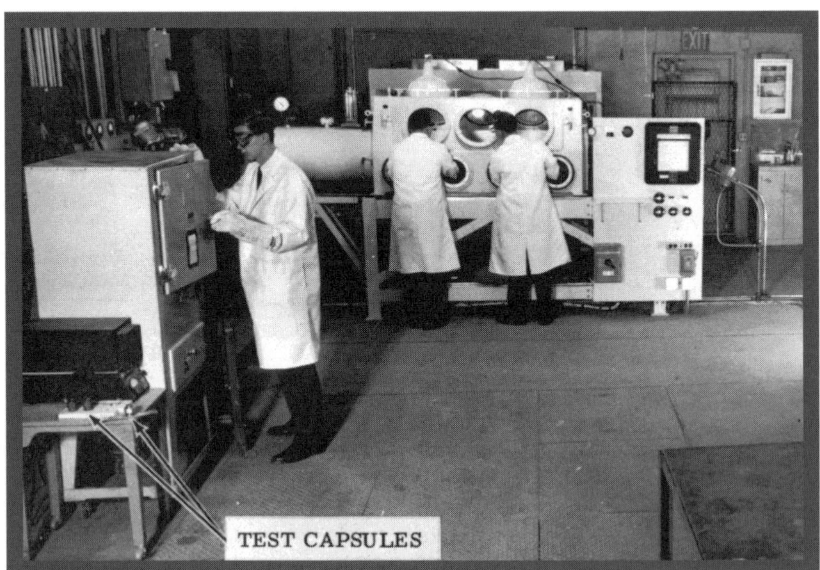

Figure 59. Personnel working at encapsulation test setup.

PROPERTIES OF LIQUID METALS		
Components of alloy	NaK-77 potassium (77.2%) sodium (22.8%)	NaKCs cesium (72.6%) potassium (24.1%) sodium (3.1%)
Type of alloy	eutectic	eutectic
Melting point (atmospheric pressure)	9.95 F	-104.8 F
Boiling point (atmospheric pressure)	1446 F	1330 F
Specific gravity	0.871 (68 F) 0.693 (1200 F)	1.467 (77 F) 1.207 (1230 F)
Absolute viscosity, centipoises	0.62 (100 F) 0.14 (1200 F)	0.95 (85 F) 0.19 (1302 F)
Thermal conductivity, watts/cm²-°C/cm	0.238 (300 F) 0.255 (1300 F)	
Electrical resistivity, microhm-cm	41.6 (300 F) 89.3 (1300 F)	76 (300 F) 139.7 (1300 F)
Average heat capacity (78.3%K weight) temperature range of 32-1292C, cal/gm/°C	0.2163	
Vapor pressure, mm Hg	1.81×10^{-3} (441 F) 431.85 (1341 F)	757 (1319 F)
Adiabatic bulk modulus, psi	463,000 (212 F) 318,000 (1000 F)	

Figure 60. Properties of liquid metal.

Figure 61. Exploded view of pump.

Figure 62. Assembled pump.

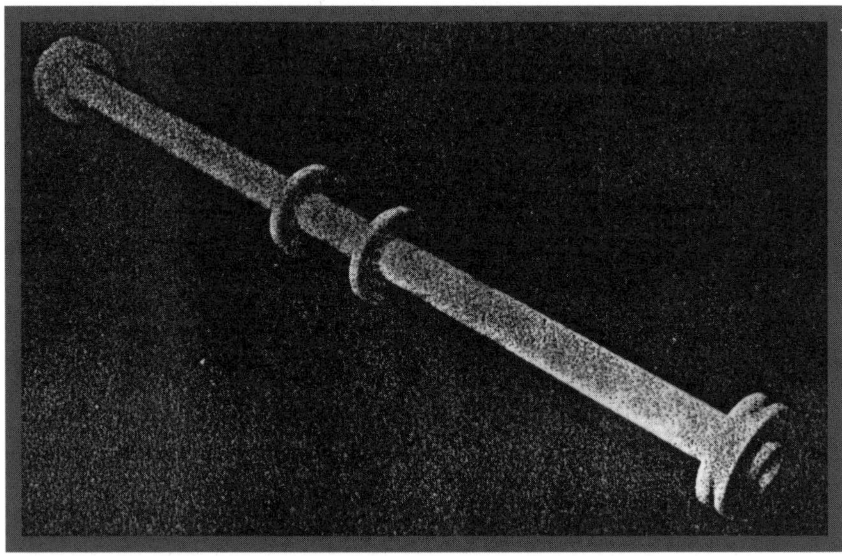

Figure 63. Alumina (Lucalox) coil bobbin.

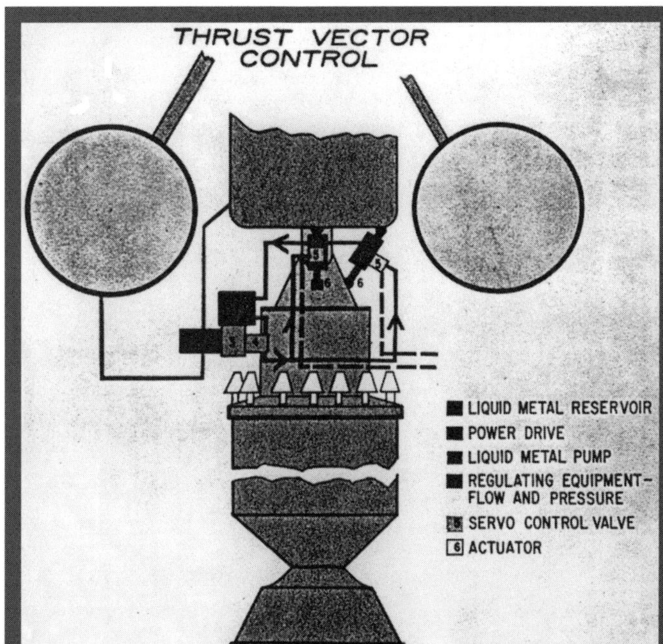

Figure 64. Thrust vector control.

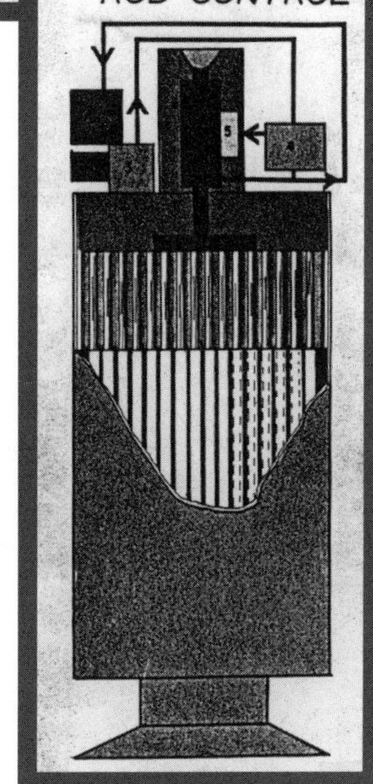

Figure 65. Fuel rod control.

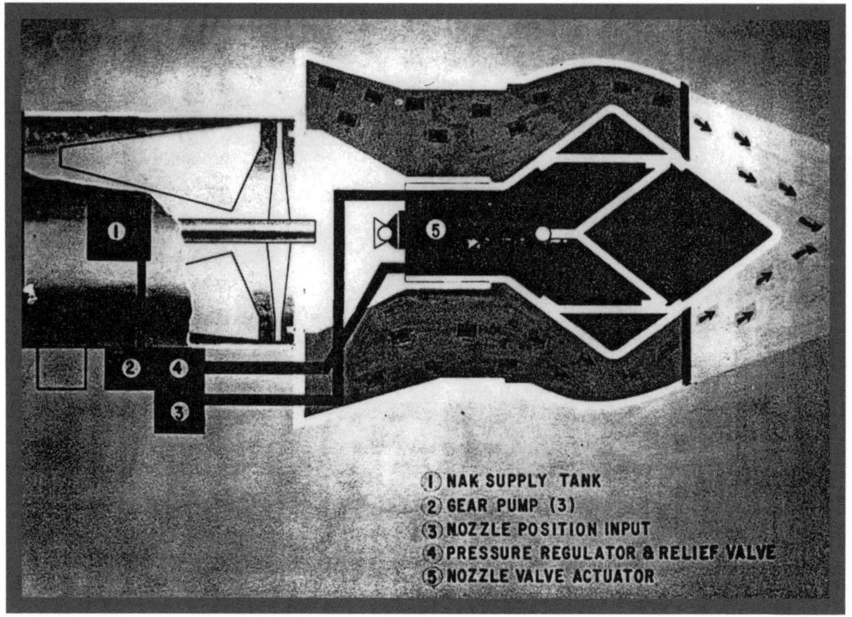

Figure 66. Air duct control.

Appendix E

Foreign Work on NaK-77

In 1967 our unit in the Flight Dynamics Laboratory at Wright Field received a proposal from the Light Military Electronics Department of the General Electric Co. to perform work on our Purchase Request on Study of Liquid Metal NaK-77, for application in flight control systems.

In essence, this was a continuation of R&D work that had started two years earlier. After several months a contract was signed and G.E. undertook the work in their proposal. One of the work items was on a liquid metal servo valve and a proposed design was submitted, shown in schematic form in Figure 67. Following is an abbreviated write-up on this design approach.

> *Proposed Study*
>
> The proposed first stage servo valve study includes the engineering analysis, design, fabrication, and test evaluation of a breadboard valve of the concept illustrated in Figure 67. Test evaluation will involve:
>
> 1. Investigation of the magnitude of variation in magnetic flux density due to changes in temperature.
> 2. Determination of the resultant variation in the magnetic flux pattern caused by changes in temperature and jet stream current.
>
> The results of this study will allow definition of efforts to reduce this concept to practical application.
>
> *Note*: Research over the next 6-8 months proved this to be a viable concept.

During the latter part of 1969, approximately one year after General Electric started work on their contract, our unit in the Flight Dynamics Laboratory received a document from the Foreign Technology Division. Figure 68 is a copy of the cover page. It should be explained here that the contents of the document are a translation of the material obtained by the Foreign Technology Division.

On page 9 of the document is a figure, noted as Fig. 2 (See Figure E-3), which is a copy of Figure 67 in the G.E. proposal and, for all practical purposes, identical. We were amazed and asked G.E. personnel how a foreign source got this data. Nobody knew.

The following statements are from this Foreign Technology Report:

> "The application of liquid metals in hydromagnetic control engineering for aircraft allows an expansion of the range of using control devices. High temperatures of even 300°C have created serious operational troubles for hydraulic devices, and pneumatic units have not always been able to match the high dynamics requirements because of the compressibility of gases."
>
> "Most frequently the eutectic sodium and potassium alloy NaK-77 is used."
>
> "Additional advantages such as radiation stability allow the use of this liquid in unit control systems of nuclear propulsion or when they come under the effect of strong radiation in cosmic space."
>
> "The application of the NaK-77 alloy described here goes beyond the bounds of experimentation. In the American program of developing air craft utilizing nuclear energy, the utilization of NaK-77 alloy as the working agent in control systems also plays an essential role, in addition to the cooling systems for reactors."

Obviously this foreign group recognized the advantages of NaK in control systems but apparently were waiting for us to solve all the hard problems of: "How does one pressurize this fluid?" and "How is it controlled?"

Even at this late date it is amazing how much others knew of our liquid metal program and how little we knew of what they knew.

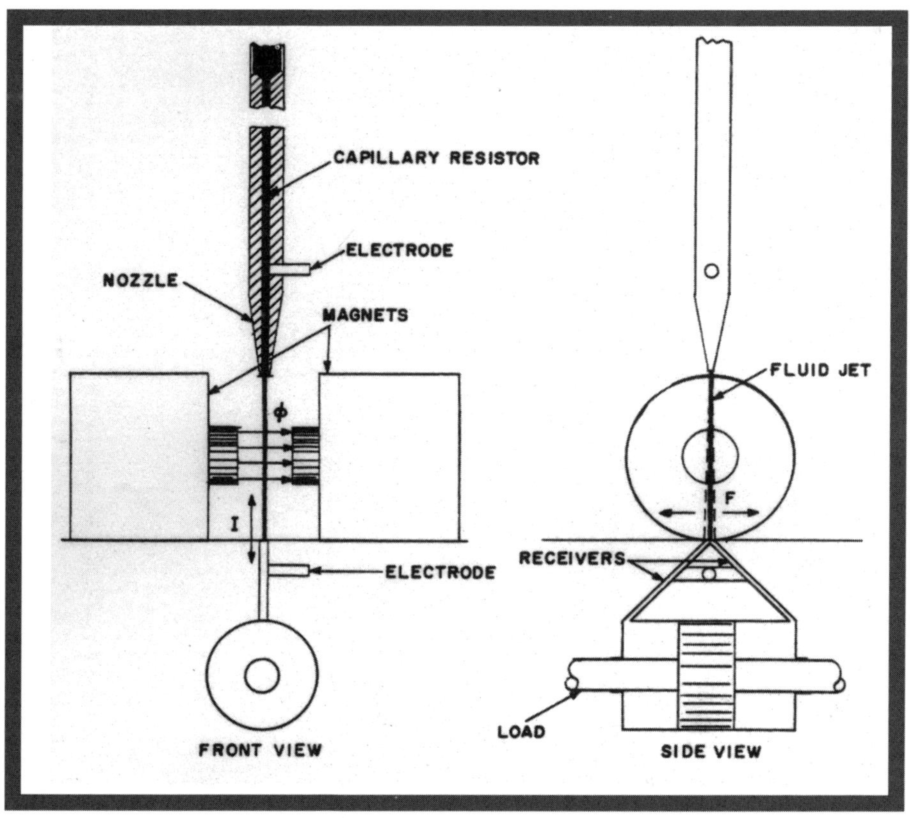

Figure 67. Electromagnetic servo valve schematic.

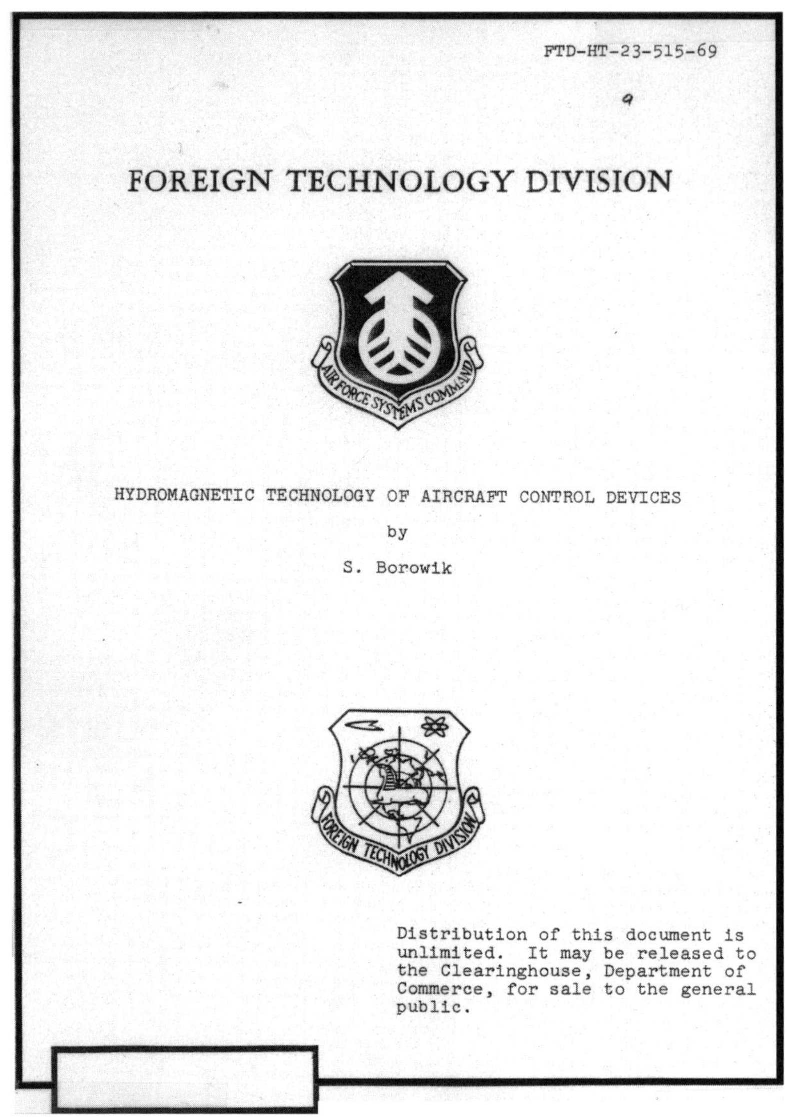

FTD-HT-23-515-69

FOREIGN TECHNOLOGY DIVISION

HYDROMAGNETIC TECHNOLOGY OF AIRCRAFT CONTROL DEVICES

by

S. Borowik

Distribution of this document is unlimited. It may be released to the Clearinghouse, Department of Commerce, for sale to the general public.

Figure 68. Cover page, document of Foreign Technology Division.

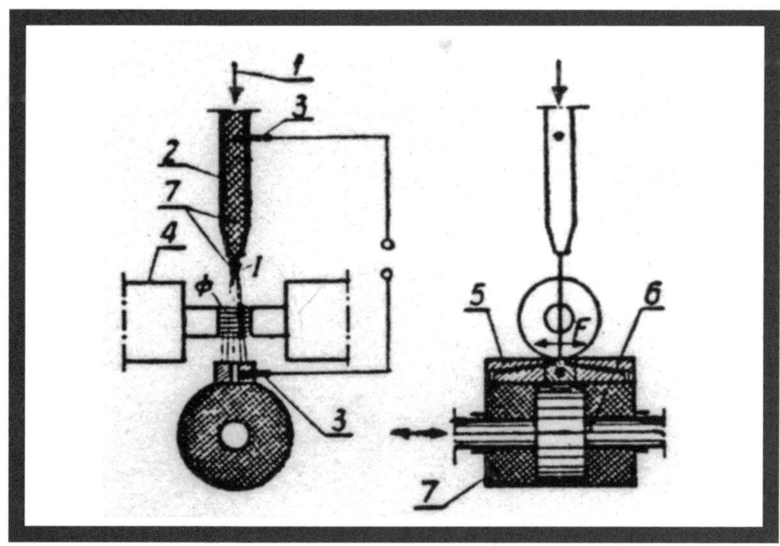

Figure 69. Figure from Foreign Technology Document.

Index

Abbreviations are used after the page number to indicate a figure (*f*).

About the Author

Vernon R. Schmitt, a native of Columbus, Ohio, is an alumnus of the Ohio State University College of Engineering. He retired as a civilian project engineer from the U.S. Air Force Flight Dynamics Laboratory at Wright-Patterson Air Force Base after spending more than 30 years on research and development programs related to flight control and servo actuation systems. After retiring, Mr. Schmitt was a consultant to the Flight Dynamics Laboratory for the next 15 years in support of its flight control research programs. He was responsible for research and development efforts on fly-by-wire by the Douglas Aircraft Company, Sperry Phoenix Corporation, LTV E-Systems, and the in-house work on the B-47 aircraft. Following World War II military service Mr. Schmitt spent two years in the Controlled Aircraft Group which also had overlapping programs on Guided Missiles. He was assigned to the Guided Missile Section at Wright-Patterson Air Force Base (at that time Wright Field) and was responsible for the flight control systems on the Matador and Rascal Missiles. These projects covered a period of about eight years. The latter part of his military career was spent with the Special Weapons Branch which dealt with controlled bombs, e.g., Azon and Razon types, and television-controlled bombs which are today referred to as "smart" bombs. Two years of his military time were spent as instructor on the Norden bombsight and the C-1 autopilot. During his instruction time Mr. Schmitt designed a controlled bomb that in many respects resembled the Razon bomb. This was submitted to the Air Force for possible use and it wasn't until the author was in the Special Weapons Branch that he realized everyone in that branch had seen and reviewed his design.